# Decentralized Estimation and Control
## for Multisensor Systems

# Decentralized Estimation and Control
## for Multisensor Systems

Arthur G.O. Mutambara

CRC Press

Boca Raton   Boston   London   New York   Washington, D.C.

**Library of Congress Cataloging-in-Publication Data**

Mutambara, Arthur G.O.
    Decentralized estimation and control for multisensor systems /
[Arthur G.O. Mutambara].
        p.  cm.
    Includes bibliographical references and index.
    ISBN 0-8493-1865-3 (alk. paper)
    1. Multisensor data fusion. 2. Automatic control. 3. Robots–
–Control systems. I. Title.
TJ211.35.M88   1998
629.8 —dc21

                                                      97-51553
                                                     CIP

# *Preface*

This book is concerned with the problem of developing scalable, decentralized estimation and control algorithms for both linear and nonlinear multisensor systems. Such algorithms have extensive applications in modular robotics and complex or large scale systems. Most existing algorithms employ some form of hierarchical or centralized structure for data gathering and processing. In contrast, in a fully decentralized system, all information is processed locally. A decentralized data fusion system consists of a network of sensor nodes, each with its own processing facility, which together do not require any central processing or central communication facility. Only node-to-node communication and local system knowledge is permitted.

Algorithms for decentralized data fusion systems based on the linear Information filter have previously been developed. These algorithms obtain decentrally exactly the same results as those obtained in a conventional centralized data fusion system. However, these algorithms are limited in requiring linear system and observation models, a fully connected sensor network topology, and a complete global system model to be maintained by each individual node in the network. These limitations mean that existing decentralized data fusion algorithms have limited scalability and are wasteful of communication and computation resources.

This book aims to remove current limitations in decentralized data fusion algorithms and further to extend the decentralized estimation principle to problems involving local control and actuation. The linear Information filter is first generalized to the problem of estimation for nonlinear systems by deriving the extended Information filter. A decentralized form of the algorithm is then developed. The problem of fully connected topologies is solved by using generalized model distribution where the nodal system involves only locally relevant states. Computational requirements are reduced by using smaller local model sizes. Internodal communication is model defined such that only nodes that need to communicate are connected. When nodes communicate they exchange only relevant information. In this way, communication is minimized both in terms of the number of communication links and size of message. The scalable network does not require propagation of information between unconnected nodes. Estimation algorithms for systems with different models at each node are developed.

The decentralized estimation algorithms are then applied to the problem of decentralized control. The control algorithms are explicitly described in terms of information. Optimal control is obtained locally using reduced order models with minimized communication requirements, in a scalable network of control nodes. A modular wheeled mobile robot is used to demonstrate the theory developed. This is a vehicle system with nonlinear kinematics and distributed means of acquiring information.

Although a specific modular robot is used to illustrate the usefulness of the algorithms, their application can be extended to many robotic systems and large scale systems. Specifically, the modular design philosophy, decentralized estimation and scalable control can be applied to the Mars Sojourner Rover with dramatic improvement of the Rover's performance, competence, reliability and survivability. The principles of decentralized multisensor fusion can also be considered for humanoid robots such as the MIT Humanoid Robot (Cog). Furthermore, the proposed decentralization paradigm is widely useful in complex and large scale systems such as air traffic control, process control of large plants, the Mir Space Station and space shuttles such as Columbia.

# *The Author*

Dr. Arthur G.O. Mutambara is an Assistant Professor of Robotics and Mechatronics in the Mechanical Engineering Department at the joint Engineering College of Florida Agricultural and Mechanical University and Florida State University in Tallahassee, Florida (U.S.A.). He has been a Visiting Research Fellow at the Massachusetts Institute of Technology (MIT) in the Astronautics and Aeronautics Department (1995), at the California Institute of Technology (CalTech) (1996) and at the National Aeronautics and Space Administration (NASA), Jet Propulsion Laboratory, in California (1994). In 1997 he was a Visiting Research Scientist at the NASA Lewis Research Center in Cleveland, Ohio. He has served on both the Robotics Review Panel and the Dynamic Systems and Controls Panel for the U.S.A. National Science Foundation (NSF).

Professor Mutambara received the Doctor of Philosophy degree in Robotics from Merton College, Oxford University (U.K.) in March 1995, where he worked with the Robotics Research Group. He went to Oxford as a Rhodes Scholar and also earned a Master of Science in Computation from the Oxford University Computing Laboratory in October 1992, where he worked with the Programming Research Group. Prior to this, he had received a Bachelor of Science with Honors in Electrical Engineering from the University of Zimbabwe in 1991.

Professor Mutambara's main research interests include multisensor fusion, decentralized estimation, decentralized control, mechatronics and modular robotics. He teaches graduate and undergraduate courses in robotics, mechatronics, control systems, estimation theory, dynamic systems and vibrations. He is a Member of the Institute of Electrical and Electronic Engineering (IEEE), the Institute of Electrical Engineering (IEE) and the British Computer Society (BCS).

# Acknowledgments

The research material covered in this book is an extension of the work I did for my Doctor of Philosophy degree at Oxford University where I worked with the Robotics Research Group. It is with great pleasure that I acknowledge the consistent and thorough supervision provided by Professor Hugh Durrant-Whyte of the Robotics Research Group, who is now Professor of Mechatronics Engineering at the University of Sydney in Australia. His resourcefulness and amazing subject expertise were a constant source of inspiration. Professor Mike Brady, Head of the Robotics Research Group at Oxford, was always accessible and supportive. My fellow graduate students in the Robotics Research Group provided the requisite team spirit and enthusiasm.

After finishing my Doctorate at Oxford University in March 1995, I took up a Visiting Research Fellowship at the Massachusetts Institute of Technology (MIT) in the Astronautics and Aeronautics Department where I carried out additional research with the Space Engineering Research Center (SERC). I would like to thank Professor Edward Crawley for inviting me to MIT and for his insightful comments. I would also like to thank Professor Rodney Brooks of the Artificial Intelligence (AI) laboratory at MIT for facilitating visits to the AI laboratory and providing information about the MIT Humanoid Robot (Cog). Further work on the book was carried out at the National Aeronautics and Space Administration (NASA) Lewis Research Center in Cleveland Ohio, where I was a Summer Faculty Research Fellow in 1997. I would like to thank Dr. Jonathan Litt of NASA Lewis for affording me that opportunity.

Quite a number of experts reviewed and appraised the material covered in this book. In particular, I would like to thank the following for their detailed remarks and suggestions: Professor Yaakov Bar-Shalom of the Electrical and Systems Engineering at University of Connecticut, Professor Peter Fleming who is Chairman of the Department of Automatic Control Engineering at the University of Sheffield (U.K.), Dr. Ron Daniel of the Robotics Research Group at the University of Oxford, Dr. Jeff Uhlmann and Dr. Simon Julier, who are both at the Naval Research Laboratory (NRL) in Washington D.C. I would also like to thank all my colleagues and students at the FAMU-FSU College of Engineering, in particular those graduate research students that I have supervised and hence unduly subjected to some of the ideas from the book: Jeff, Selekwa, Marwan, Rashan, Robert and Todd. Their questions and comments helped me make some of the material more readable.

I attended Oxford University as a Rhodes Scholar and visited robotics research laboratories (both academia and industry) in the United States of America, Japan, Germany and the United Kingdom while presenting papers at international conferences, courtesy of funds provided by the Rhodes Trust. Consequently, financial acknowledgment goes to the generality of the struggling people of Southern Africa who are the living victims of the imperialist Cecil John Rhodes. Every Rhodes Scholar should feel a sense of obligation and duty to the struggle of the victims of slavery, colonialism and imperialism throughout the world.

*This book is dedicated to oppressed people throughout the world and their struggle for social justice and egalitarianism. Defeat is not on the agenda.*

# Contents

# Chapter 1

## Introduction

### 1.1 Background

This book is concerned with the problem of developing scalable decentralized estimation and control algorithms for both linear and nonlinear multisensor systems.

A sensor is any device which receives a signal or stimulus and generates measurements that are functions of that stimulus. Sensors are used to monitor the operation of a system and to provide information through which the system may be controlled. In this way, a sensor allows a system to learn and continuously update its own model of the world. However, a single sensor is not always capable of obtaining all the required information reliably at all times in varying environments. Furthermore, as the complexity of a system increases so does the number and variety of sensors required to provide a complete description of the system and allow for its effective control. Multiple sensors provide a better and more precise understanding of the system and its operation. Multisensor systems have found wide applications in areas such as robotics, aerospace, defense, manufacturing, process control and power generation.

A multisensor system may employ a range of different sensors, with different characteristics, to obtain information about an environment. The diverse and sometimes conflicting information obtained from multiple sensors gives rise to the problem of how the information may be combined in a consistent and coherent manner. This is the *data fusion* problem. Multisensor fusion is the process by which information from a multitude of sensors is combined to yield a coherent description of the system under observation. Both *quantitative* and *qualitative* sensor fusion methods have been advanced in the literature. Quantitative methods are used exclusively in this book. They are based on probabilistic and statistical methods of modeling and combining information. Quantitative techniques include methods of statistical decision theory, Bayesian analysis and filtering techniques.

*Kalman* filtering and its algebraically equivalent technique, *information* filtering, are quantitative data fusion methods based on linear decision rules. The Information filter essentially tracks information about states and not the states themselves. The properties of information variables enable this filter to be easily distributed and decentralized. The work described in this book is based on these methods.

A variety of information based data fusion algorithms have been employed in recent work [22], [41], [71], [80]. In this work extensive descriptions of *centralized, hierarchical* and *decentralized* architectures and their advantages and limitations are discussed. Emphasis is placed on fully decentralized sensing based on the linear Information filter. A fully decentralized system is defined as a data processing system in which all information is processed locally and there is no central processing site. It consists of a network of sensor nodes, each with its own processing facility, which together do not require any central fusion or communication facility. Special Transputer based architectures have been built to demonstrate that the principle of decentralized sensing is indeed viable. Elsewhere, research work using conventional state space multisensor fusion methods has also been extensive, as evidenced by the work of Abidi and Gonzalez [1], Aggarwal [3], Bar-Shalom [14], Luo [68], McKendall and Mintz [77], and Richard and Marsh [111].

Most of the current sensor fusion algorithms consider systems described by linear dynamics and observation models. Most practical problems have nonlinear dynamics and sensor information nonlinearly dependent on the states that describe the environment. Although, linearization methods such as the extended Kalman filter are popular, there is currently no algorithm that solves the nonlinear data fusion problem in Information filter form. Given the advantages of using information variables in distributed and decentralized fusion, this is an extremely important case to address. Another major drawback of the algorithms presented to date is that although they tell us how to fuse information, they do not say how to use this fused information to *control* the system. The applications of decentralized multisensor and multiactuator control are potentially huge. Research on systems that have been described as '*decentralized*' control has been prolific. The definition of a decentralized system has been varied, in some cases simply referring to schemes involving more than one controller. Work in this field has included that of Chong and Mori [38], Hashemipour [50], Sandell [113], Siljak [115] and Speyer [117]. The issue, however, is that most of these systems are not fully decentralized and they do not exploit the use of information variables. In these systems, some central processing site is always retained, leading to an essentially hierarchical structure consisting of interacting levels.

The work by Speyer is the exception. However, he does not exploit the use of information variables. Moreover, in Speyer's algorithm and fully decentralized estimation algorithms in [41], [71], [106], the sensing network topology is fully connected, that is, each local sensing node communicates with all the other nodes. This poses serious problems of communication redundancy, duplication of computation and limited system scalability. Furthermore, loss of any one communication link violates the fully connected assumption. In fully connected networks local models of state, information and control are the same as those of the equivalent centralized system. Consequently, the decentralized control algorithm derived from such a network is essentially the centralized controller repeated at each node. This is of limited practical benefit, particularly for a large system with a large number of nodes.

There have been efforts to derive non-fully connected decentralized estimation topologies [48], [54], using a special internodal filter, the *channel filter*. This is an additional filter which integrates information common to two communicating nodes. It is used to propagate information between two unconnected nodes. Interesting though this approach is, it still employs the same size of variables locally as in the centralized case and the additional filtering process at each node increases the computational load. Moreover, this work only addresses estimation in linear systems and not nonlinear estimation or control systems.

## 1.2 Motivation

The motivation for the material presented in this book derives from two aspects of the work discussed above. The first point of motivation is the benefits of multisensor systems, in particular decentralized methods of data fusion and control. The second point derives from the limitations of existing decentralized methods. This book seeks to develop fully decentralized data fusion and control algorithms which do not exhibit the drawbacks of existing methods. In particular, it aims to address both the problem of using reduced local models at sensor nodes and that of reducing communication and connection requirements.

The estimation and control algorithms developed have potential applications in multisensor systems and large scale systems, which are also often characterized by multiple actuators, controllers, targets and trackers. The algorithms can also be applied in such fields as space structures and flexible structures.

## 1.2.1  Modular Robotics

Of the many multisensor systems that motivate the theory developed in this book, modular robotics is the most specific application of interest. A modular vehicle has the same function as any conventional robot except that it is constructed from a small number of standard units. Each module has its own hardware and software, driven and steered units, sensors, communication links, power unit, kinematics, path planning, obstacle avoidance, sensor fusion and control systems. There is no central processor on the vehicle. Vehicle kinematics and dynamics are invariably nonlinear and sensor observations are also not linearly dependent on the sensed environmental states. These kinematics, models and observation spaces must be distributed to the vehicle modules.

The vehicle employs multiple sensors to measure its body position and orientation, wheel positions and velocities, obstacle locations and changes in the terrain. Sensor information from the modules is fused in a decentralized way and used to generate local control for each module. The advantages of this modular technology include reduction of system costs, application flexibility, system reliability, scalability and survivability. However, for the modularity to be functional and effective, fully decentralized and scalable multisensor fusion and control are mandatory.

## 1.2.2  The Mars Sojourner Rover

One robotic vehicle that has recently fired many researchers' imagination is the NASA Mars Pathfinder Mission's Sojourner Rover which is currently carrying out exploration on Mars. The Prospector spacecraft containing the Rover landed on Mars on July 4th 1997. The Mars Pathfinder Rover team plans a vehicle traverse from the Rover Control Workstation at NASA (Jet Propulsion Laboratory) in Pasadena, California. Due to the speed of light time delay from Earth to Mars (11 minutes), and the constraint of a single uplink opportunity per day, the Rover is required to perform its daily operations autonomously. These activities include terrain navigation, rock inspection, terrain mapping and response to contingencies [43].

During traverses the Rover uses its look ahead sensors (5 laser stripe projectors and two CCD cameras) to detect and avoid rocks, dangerous slopes and drop off hazards, changing its path as needed before turning back towards its goal. Bumpers, articulation sensors and accelerometers allow the Rover to recognize other unsafe conditions. The hazard detection system can also be adopted to center the Rover on a target rock in preparation for deployment of its spectrometer. Other on-board experiments characterize soil mechanics, dust adherence, soil abrasiveness and vehicle traverse performance. A picture of the Mars Rover is shown in Figure 1.1.

**FIGURE 1.1**
**The Mars Sojourner Rover: A Multisensor System. (Photo Courtesy of NASA)**

The capability of the Rover to operate in unmodeled environment, choosing actions in response to sensor inputs to accomplish requested objectives, is unique among robotic space missions to date. Being such a complex and dynamic robotic vehicle characterized by a myriad of functions and different types of sensors while operating in an unmodeled and cluttered environment, the Sojourner Rover is an excellent example of a multisensor and multiactuator system. Establishing efficient and effective multisensor fusion and control for such a system provides motivation for the material presented in this book. How can the vehicle combine and integrate information obtained from its multiple sensors? How can it optimally and efficiently use this information to control its motion and accomplish its tasks, that is, achieve intelligent connection of perception to action? Currently the principal sensor fusion algorithms being used on the Rover are based on state space methods, in particular, the extended Kalman filter; and these data fusion algorithms and their corresponding architectures are ostensibly centralized [73]. There is also very little modularity in the hardware and software design of the Rover.

Design modularity, decentralized estimation and control provide certain advantages that would be relevant to the Rover. For example, if each wheel or unit is monitored and controlled by an independent mechanism, then decentralized sensor processing and local control can permit the Rover to continue its mission even if one or more wheels/units are incapacitated.

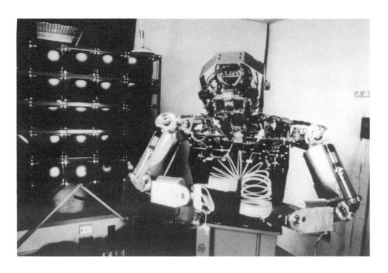

**FIGURE 1.2**
**The MIT Humanoid Robot (Cog): A Multisensor System.**
**(Photo Courtesy of Donna Coveney/MIT)**

In addition, information from the various sensors will be efficiently utilized, thus, optimally taking advantage of the redundancy inherent in the Rover's multiple sensors. It is submitted here that if the estimation, control and design paradigm proposed in this book is adopted for the Mars Sojourner Rover, its competence, reliability and survivability could be improved.

## 1.2.3   The MIT Humanoid Robot (Cog)

The principle behind creating the MIT Humanoid Robot (Cog) derives from the hypothesis that humanoid intelligence requires humanoid interactions with the world. The form of the human body is critical to the representations that are developed and used for both human internal thought and language. If a robot with human-like intelligence is to be built, then it must have a human-like body in order to be able to develop similar representations. A second reason for building a humanoid form is that an important aspect of being human is interaction with other humans. For a human-level intelligent robot to gain experience in interacting with humans it needs a large number of interactions. If the robot has humanoid form then it will be both easy and natural for humans to interact with it in a human-like way. In this way a large source of dynamic interactions is obtained which will not be possible with disembodied human intelligence. Hence, in order to understand human cognition and utilize it in machines, it is necessary to built a humanoid robot [31], [32].

The entire mission of building a humanoid robot would be inconceivable without the use of multiple sensors. MIT's Cog is a set of multiple sensors and multiple actuators which approximate the sensory and motor dynamics of a human body. The sensory functions include sight (video cameras), hearing, touch, proprioception (joint position and torque), a vestibular system and a vocalization system. Cog's "brain" is a large scale MIMD (multiple input and multiple data) computer architecture which consists of a set of Motorolla 68332 processors executing parallel computations. Its head and visual system is designed such that it approximates the complexities of the human visual system and the output is displayed on the rack of twenty monitors. Cog's eye, the camera system, has four degrees of freedom consisting of two active "eyes". To mimic human eye movements, each eye can rotate about a vertical axis and a horizontal axis [72], [127].

With such a myriad of multiple sensors in the humanoid robot, it is essential that the issue of multisensor fusion is appropriately addressed so that the information from the sensors is efficiently and optimally used.

### 1.2.4 Large Scale Systems

The problems of monitoring, supervising and controlling large scale systems also provide a compelling case for the material presented in this book. A large scale system is defined as a group of subsystems that are interconnected in such a way that decentralized operation is mandatory. Such systems have a large number of sensors and actuators, and a large dimensionality (i.e., a large number of states). A large scale system is so physically dispersed such that a centralized sensor fusion center or controller would be prohibitively expensive. Furthermore, sometimes the system is known to be weakly coupled so that the degradation in performance resulting from forced decentralization should be modest.

Systems that can be classified as large scale include the following: an urban traffic control system, control of a large paper making plant, an air traffic control system, control of a large processing plant and a military command control system. Two examples of large scale systems which are also complex are the Space Shuttle Columbia and the Russian Mir Station. Their main features and functions are described in the next subsections in order to capture the complexity and extensiveness of such systems, thus amply illustrating the case for both decentralized multisensor fusion and decentralized control.

### 1.2.5 The Russian Mir Space Station

The Russian Mir Space Station which was launched into space in February 1986 has been in orbit for eleven years, and staffed continuously for the past six years. It consists of modules launched separately and brought

**FIGURE 1.3**
**The Mir Station: A Complex and Large Scale System. (Russian Space Agency Photo Courtesy of NASA)**

together in space, and it weighs more than one hundred tons. The design philosophy behind the Mir station is that of an assembly of separate pressurized modules with both core and specialized functions. As of November 1997, the modular station consists of the Mir core, Kvant 1, Kvant 2, Kristall, Spektr, Priroda and Docking modules [100]. Mir measures more than 107 feet long and is about 90 feet wide across its modules. A picture of the station in space is shown in Figure 1.3.

The 20.4 ton Core module is the central portion and the first building block of the Mir station which supports the modular design. It provides basic services (living quarters, life support, power) and scientific research capabilities. Kvant 1 is a small, 11-ton module which contains astrophysics instruments, life support and altitude control equipment. The purpose of the Kvant-1 module is to provide data and observations for research into the physics of active galaxies, quasars, and neutron stars. The Kvant-2 module which weighs 19.6 tons carries an EVA airlock, solar arrays, life support equipment, drinking water, oxygen provisions, motion control systems, power distribution and washing facilities. Its purpose is to provide biological research data, earth observation data and EVA capability [100].

The Spektr Module carries four solar arrays and scientific equipment, and its main focus is earth observation, specifically natural resources and atmosphere. Kristall module carries scientific equipment, retractable solar arrays, and a docking node equipped with a special androgynous docking mechanism designed to receive a spacecraft weighing up to 100 tons. The Docking module allows a space shuttle to dock with the Mir station without interfering with the solar arrays. The purpose of the Kristall module is to develop biological and materials production technologies in the space environment. Priroda module's primary function is to add earth remote-sensing capability to Mir and contains the hardware and supplies for several joint U.S.-Russian science experiments. Its earth remote-sensing capabilities include, monitoring the ecology of large industrial areas, measuring concentration of gaseous components in the atmosphere, determining temperature fields on the ocean surface, and monitoring the process of energy and mass exchange between ocean and atmosphere which affect the weather.

Clearly, the Mir station is a large, modular and dispersed system which employs a huge number of sensors, actuators and controllers to carry out the functions of its various modules. It is inconceivable and impractical to consider centralized multisensor fusion or centralized control for such a system.

## 1.2.6   The Space Shuttle Columbia

The space shuttle Columbia, also referred to as Orbiter Vehicle-102, is the oldest orbiter in the shuttle fleet and was the first U.S.A. space shuttle to fly into earth orbit in 1981. Over the years it has been updated and modified several times. It has carried out 23 flights and 3,286 orbits, and has spent a total of 196 days in space [98], [99]. Since 1981 four other ships have joined the fleet; Challenger in 1982 (but destroyed four years later), Discovery in 1983, Atlantis in 1985 and Endeavor which was built as a replacement for Challenger in 1991. The last shuttle mission of 1997, the Space Shuttle Columbia STS-87, was launched into space on the 19th of November from the Kennedy Space Center in Florida, U.S.A. Figure 1.4 shows a picture of Columbia blasting off the launch pad into space. In order to illustrate the complexity of a space shuttle and show the diversity and multiplicity of its sensors, some of the experiments and instrumentation on the Columbia STS-87 mission are briefly described here.

The objective of the mission is to carry out several scientific experiments in space. The United States Microgravity Payload (USMP) is a spacelab consisting of microgravity research experiments, while the Solar Physics Spacecraft (SPS) is to perform remote-sensing of the hot outer layers of the sun's atmosphere. The Space Acceleration Measurement System (SAMS) is a microprocessor-driven data acquisition system designed to measure and record the microgravity acceleration environment of the USMP carrier.

**FIGURE 1.4**
**The Space Shuttle Columbia:  A Complex and Large Scale**
**System. (Photo Courtesy of NASA)**

The Orbital Acceleration Research Experiment (OARE) is a highly sensitive instrument designed to acquire and record data of low-level aerodynamic acceleration along the orbiter's principal axes in the free-molecular flow regime at orbital altitudes [99].

The objective of the Shuttle Ozone Limb Sounding Experiment (SOLSE) is to determine the altitude distribution of ozone in an attempt to understand its behavior so that quantitative changes in the composition of the atmosphere can be predicted, whereas the Loop Heat Pipe (LHP) test advances thermal energy management technology and validates technology readiness for upcoming commercial spacecraft applications. The Sodium Sulfur Battery Experiment (NaSBE) characterizes the performance of four 40 amp-hour sodium-sulfur battery cells. In order to gain an understanding of the fundamental characteristics of transitional and turbulent gas jet diffusion flames under microgravity conditions, the Turbulent Gas Jet Diffusion (G-744) experiment is provided. The Autonomous EVA Robotic Camera (AERC) is a small, unobtrusive, free-flying camera platform for use outside a spacecraft. On board the free-flyer are rate sensors to provide data for an automatic altitude hold capability.

The Shuttle Infrared Leeside Temperature Sensing (SILTS) experiment is used to obtain high-resolution infrared imagery of the upper (leeward) surface of the orbiter fuselage and left wing during atmospheric entry. This information is hoped to increase understanding of leeside aeroheating phenomena and can be used to design a less conservative thermal protection system. The primary components of the SILTS system include an infrared camera, infrared-transparent windows, a data and control electronics module, and a pressurized nitrogen module. Accurate aerodynamic research requires precise knowledge of vehicle altitude and state. This information, commonly referred to as air data, includes vehicle angle of attack, angle of sideslip, free-stream dynamic pressure, Mach number and total pressure. Hence the Shuttle Entry Air Data System (SEADS) was developed to take the measurements required for precise determination of air data across the orbiter's atmospheric flight-speed range.

The Shuttle Upper Atmosphere Mass Spectrometer (SUMS) experiment is for obtaining measurements of free-stream density during atmospheric entry in the hypersonic, rarefied flow regime. These measurements, combined with acceleration measurements from the companion high-resolution accelerometer package experiment, allow calculation of orbiter aerodynamic coefficients in the flow regime previously inaccessible using experimental and analytic techniques. The High Resolution Accelerometer Package (HRAP) experiment uses an orthogonal, triaxial set of sensitive linear accelerometers to take accurate measurements of low-level (down to micro-gs) aerodynamic accelerations along the orbiter's principal axes during initial re-entry into the atmosphere, that is, in the rarefied flow regime.

The Orbiter operational instrumentation (OI) is used to collect, route and process information from transducers and sensors throughout the orbiter and its payloads. This system also interfaces with the solid rocket boosters, external tank and ground support equipment. The instrumentation system consists of transducers, signal conditioners, two pulse code modulation master units, encoding equipment, two operational recorders, one payload recorder, master timing equipment and on-board checkout equipment. The OI system senses, acquires, conditions, digitizes, formats and distributes data for display, telemetry, recording and checkout. The digital signal conditioners convert digital and analog data signals from the various sensors into usable forms. These measured parameters include frequency, voltage, current, pressure, temperature (variable resistance and thermocouple), displacement (potentiometer) [98].

The Network Signal Processor (NSP) is the nucleus of the communication systems and it is responsible for processing and routing commands, telemetry and voice between the orbiter and the ground. The Closed-Circuit Television System (CCTV) is used primarily to support on-orbit activities that require visual feedback to the crew. The CCTV system also provides the capability to document on-orbit activities and vehicle configurations for permanent record or for real-time transmission to the ground. Typical uses of the CCTV monitoring system include payload bay door operations, remote manipulator system operations, experiment operations, rendezvous and station keeping operations, and various on-board crew activities [99].

The CCTV system consists of the video control unit, television cameras, VTR and two on-board television monitors.

From the above descriptions of shuttle experiments and instrumentation it is evident that there is need for decentralized and synergistic integration of information from sensors in addition to decentralized supervision and control of the different shuttle units.

## 1.3   Problem Statement

The problem addressed in this book is that of formulating algorithms which obtain globally *optimal* state estimates and control locally, subject to the following constraints:

- No node acts as a central processing site for fusion or control, and the size of the system and number of nodes are arbitrary.

- Only nodes with a common state space, observed by either or both nodes, communicate. Any such communicating nodes exchange only relevant information and there is no propagation of information between any two unconnected nodes.

- Only locally relevant computation takes place, thus reducing local computational requirements.

- The observation space and system dynamics are nonlinear.

Optimal here means the estimate or control signal at each node is equivalent to that obtained by a corresponding centralized system. Optimality concepts are traditionally asserted in the context of centralized systems, where the optimization criterion for an estimator is usually the minimization of the covariance while, for control, it is minimization of a performance criterion.

In terms of applications the specific motivation is the design of a decentralized sensor fusion and control system for a modular wheeled mobile robot. This is a robot vehicle system with nonlinear kinematics and with distributed sources of sensor information.

## 1.4 Approach

### 1.4.1 Estimation

The approach adopted is to extend the algebraic equivalent of the Kalman filter, the Information filter to problems involving both system and observation nonlinearities. The data fusion problem in nonlinear multisensor systems is then considered and a decentralized linearized estimation algorithm proposed. Considering problems of full connectedness leads to the use of model distribution methods, where local models involve only relevant global states. In such a system, communication is achieved by model defined internodal communication. Estimation algorithms for nodes using reduced order models are thus derived.

### 1.4.2 Control

The key issue is identified as complementarity between data fusion and control. This is because two distinct but complementary theories of *data fusion* and *control* are required to solve the problem stated above. It then becomes pertinent to understand the relationship between estimation and sensor based control. The central organizing principle in this book is the separation of estimation from control. The two are solved as separate but complementary subproblems. For linear systems this is justified by the *separation* and *certainty equivalence* principles. In the nonlinear case, the notion of *assumed certainty equivalence* is employed. In both cases an optimal estimator, separately designed, is cascaded with the corresponding optimal deterministic feedback control gain. Optimal stochastic control for a linear, quadratic and Gaussian *(LQG)* problem is considered. The optimal deterministic control gain is generated from *backward Riccati* recursion using the optimality principle and stochastic dynamic programming. Expressing the control law in terms of information estimates, an information form of the standard LQG controller is derived. A system with several actuators is then configured into a fully connected topology of decentralized communicating control nodes. Control vectors, models and information vectors are then distributed to resolve issues of full connectedness.

### 1.4.3 Applications

The proposed theory is tested using software written in parallel *ANSI C* running on Transputer based parallel hardware. Some demonstrative simulations are run using Matlab. Validation is carried out by comparing the results of the distributed and decentralized systems with corresponding conventional centralized controllers. Application of the theory to control a modular Wheeled Modular Robot (**WMR**) is demonstrated. This is done by distributing the vehicle kinematics, constructing a vehicle model and then developing generic software which is the same at each vehicle module. Test runs are carried out for a number of WMR trajectories. The principal goal is to demonstrate the effectiveness of decentralized WMR estimation and control.

## 1.5   Principal Contributions

This book makes a number of theoretical and practical contributions in the area of decentralized estimation and control for multisensor systems and large scale systems:

- The linear Information filter is generalized and extended to the problem of estimation for nonlinear systems by deriving the extended Information filter (**EIF**). A decentralized form of the algorithm, the decentralized extended Information filter (**DEIF**), is also developed, thus, generalizing methods traditionally applied for decentralized estimation in linear systems to the much larger class of applications involving nonlinear systems.

- Solutions to the generalized model distribution problem in decentralized data fusion and control systems are presented. This allows for model defined, non-fully connected estimation and control networks based on internodal information transformation. In these topologies there is local internodal communication and no propagation of information between unconnected nodes. The main advantages of these networks are reduced computation and minimized communication.

- Estimation algorithms for systems with different models at each node are derived. For linear systems, the distributed and decentralized Information filter (**DDIF**) is developed and for nonlinear systems the distributed and decentralized extended Information filter (**DDEIF**) is developed.

- Fully decentralized estimation algorithms are applied to the problem of decentralized control for both linear and nonlinear systems. The control algorithms are explicitly expressed in terms of information. Globally optimal control is obtained locally using reduced order models with minimized communication requirements.

- A decentralized kinematic model and modular software for any wheeled mobile robot (**WMR**) with *simple* wheels is contributed. Generic software based on Transputer technology is developed which can be loaded onto a vehicle of any kinematic configuration.

- The internodal transformation theory provides a formal WMR design methodology by specifying which vehicle modules have to communicate and what information they have to exchange. In this way scalable and efficient WMR configurations can be derived.

The value of the extended Information filter is further enhanced by its flexibility to work with recently developed techniques for improving the accuracy and generality of Kalman and extended Kalman filters. Specifically, the Unscented Transform provides a mechanism for applying nonlinear transformations to the mean and covariance estimates that is provably more accurate than standard linearization [60], [62], [105]. The EIF can also be extended to exploit the generality of Covariance Intersection (CI) to remove the independence assumptions required by all Kalman-type update equations [61], [123], [124]. All results relating to the EIF can be easily extended to exploit the benefits of the Unscented Transform and CI.

## 1.6   Book Outline

The current chapter, (**Chapter 1**), provides the background and motivation for the work covered. In **Chapter 2** the essential estimation techniques used in this book are introduced. These techniques are based on the Kalman filter, a widely used data fusion algorithm. The Information filter, an algebraic equivalent of the Kalman filter, is derived. The advantage of this filter in multisensor problems is discussed. For nonlinear systems the conventional extended Kalman filter is derived. For use in multisensor problems, involving nonlinearities, the extended Information filter is developed by integrating principles from the extended Kalman and linear Information filters. Examples of estimation in linear and nonlinear systems are used to validate the Information filter and EIF algorithms with respect to those of the Kalman filter and EKF.

**Chapter 3** extends the estimation algorithms of Chapter 2 to fully connected, multisensor decentralized estimation problems. An overview of multisensor systems, fusion architectures and data fusion methods is given. A definition of a decentralized system is given and the literature in this area is discussed. Decentralized estimation schemes consisting of communicating sensor nodes are then developed by partitioning and decentralizing the state and information space filters of Chapter 2. In this way four decentralized estimation algorithms are derived and compared. The decentralized extended Information filter (DEIF) is a new result which serves to address the practical constraint of system nonlinearities. However, all four of the decentralized estimation algorithms developed require *fully connected* networks of communicating sensor nodes in order to produce the same estimates as their corresponding centralized systems. The problems associated with fully connected decentralized systems are discussed.

In **Chapter 4** the problems arising from the constraint of fully connectedness are resolved by removing it. This is accomplished by using distributed reduced order models at local nodes, where each local state consists only of locally relevant states. Information is exchanged by model defined, internodal communication. Generalized internodal transformation theory is developed for both state space and information space estimators. The network topology resulting from this work is model defined and non-fully connected. Any two unconnected nodes do not have any relevant information for each other, hence there is no need to propagate information between them. Scalable decentralized estimation algorithms for non-fully connected topologies are then derived for both linear and nonlinear systems. The most useful algorithm is the distributed and decentralized extended information filter (DDEIF). It provides scalable, model distributed, decentralized (linearized) estimation for nonlinear systems in terms of information variables.

In **Chapter 5** the decentralized estimation algorithms from Chapter 3 and 4 are extended to the problem of decentralized control. First, for a single sensor-actuator system the standard stochastic LQG control problem is solved using information variables. The same approach is used for the (linearized) nonlinear stochastic control problem. Equipped with the information forms of the LQG controller and its nonlinear version, decentralized multisensor and multiactuator control systems are then considered. A decentralized algorithm for a fully connected topology of communicating control nodes is derived from the estimation algorithms of Chapter 3. The attributes of such a system are discussed. By removing the constraint of fully connectedness as discussed in Chapter 4, the problem of scalable decentralized control is developed. Of most practical value is the *distributed* and decentralized control algorithm, expressed explicitly in terms of information, which applies to systems with nonlinearities. The advantages of model defined, non-fully connected control systems are then presented. Simulation examples are also presented.

In **Chapter 6** the hardware and software implementation of the theory is described. A general decentralized and modular kinematic model is developed for a WMR with simple wheels. This is combined with the decentralized control system from Chapter 5 to provide a modular decentralized WMR control system. The actual WMR system models used in the implementation are presented. The modular vehicle used in this work is briefly introduced and the units of the WMR described. Examples of complete assembled vehicle systems are presented to illustrate design flexibility and scalability. The Transputer based software developed is then outlined and explained using pseudocode. Software modularity is achieved by using a unique configuration file and a generic nodal program.

In **Chapter 7** the experimental results are presented and analyzed. The key objective is to show that given a good centralized estimation or control algorithm, an equally good decentralized equivalent can be provided. This is done by using results from both simulations and the WMR application. The same performance criteria are used for centralized and decentralized systems. For estimation, the innovations sequences are analyzed, while the estimated control errors (from reference trajectories) are used to evaluate control performance. The results are discussed and conclusions drawn.

In **Chapter 8** the work described in the book is summarized and future research directions explored. First, the contributions made are summarized and their importance put into the context of existing decentralized estimation and control methods. The limitations of the techniques developed are identified and possible solutions advanced. Research fields and applications to which the work can be extended are proposed.

# Chapter 2

## Estimation and Information Space

### 2.1 Introduction

In this chapter the principles and concepts of estimation used in this book are introduced. General recursive estimation and, in particular, the *Kalman filter* is discussed. A Bayesian approach to probabilistic information fusion is outlined. The notion and measures of information are defined. This leads to the derivation of the algebraic equivalent of the Kalman filter, the (linear) *Information filter*. The characteristics of this filter and the advantages of information space estimation are discussed. State estimation for systems with nonlinearities is considered and the *extended* Kalman filter treated. Linear information space is then extended to *nonlinear* information space by deriving the *extended* Information filter. This establishes all the necessary mathematical tools required for exhaustive information space estimation. The advantages of the extended Information filter over the extended Kalman filter are presented and demonstrated. This filter constitutes an original contribution to estimation theory and forms the basis of the decentralized estimation and control methods developed in this book.

### 2.2 The Kalman Filter

All data fusion problems involve an estimation process. An estimator is a decision rule which takes as an argument a sequence of observations and computes a value for the parameter or state of interest. The *Kalman filter* is a recursive linear estimator which successively calculates a minimum variance estimate for a state, that evolves over time, on the basis of periodic observations that are linearly related to this state. The Kalman filter estimator minimizes the mean squared estimation error and is optimal with respect to a variety of important criteria under specific assumptions about process and observation noise. The development of linear estimators can be extended to the problem of estimation for nonlinear systems. The Kalman filter has found extensive applications in such fields as aerospace navigation, robotics and process control.

## 2.2.1   System Description

A very specific notation is adopted to describe systems throughout this book [12]. The state of nature is described by an $n$-dimensional vector $\mathbf{x}=[x_1, x_2, ..., x_n]^T$. Measurements or observations are made of the state of $\mathbf{x}$. These are described by an $m$-dimensional observation vector $\mathbf{z}$.

A linear discrete time system is described as follows:

$$\mathbf{x}(k) = \mathbf{F}(k)\mathbf{x}(k-1) + \mathbf{B}(k)\mathbf{u}(k-1) + \mathbf{w}(k-1), \qquad (2.1)$$

where $\mathbf{x}(k)$ is the state of interest at time $k$, $\mathbf{F}(k)$ is the state transition matrix from time $(k-1)$ to $k$, while $\mathbf{u}(k)$ and $\mathbf{B}(k)$ are the input control vector and matrix, respectively. The vector, $\mathbf{w}(k) \sim N(\mathbf{0}, \mathbf{Q}(k))$ is the associated process noise modeled as an uncorrelated, zero mean, white sequence with process noise covariance,

$$\mathrm{E}[\mathbf{w}(i)\mathbf{w}^T(j)] = \delta_{ij}\mathbf{Q}(i).$$

The system is observed according to the linear discrete equation

$$\mathbf{z}(k) = \mathbf{H}(k)\mathbf{x}(k) + \mathbf{v}(k), \qquad (2.2)$$

where $\mathbf{z}(k)$ is the vector of observations made at time $k$. $\mathbf{H}(k)$ is the observation matrix or model and $\mathbf{v}(k) \sim N(\mathbf{0}, \mathbf{R}(k))$ is the associated observation noise modeled as an uncorrelated white sequence with measurement noise covariance,

$$\mathrm{E}[\mathbf{v}(i)\mathbf{v}^T(j)] = \delta_{ij}\mathbf{R}(i).$$

It is assumed that the process and observation noises are uncorrelated, i.e.,

$$\mathrm{E}[\mathbf{v}(i)\mathbf{w}^T(j)] = \mathbf{0}.$$

The notation due to Barshalom [12] is used to denote the estimate of the state $\mathbf{x}(j)$ at time $i$ given information up to and including time $j$ by

$$\hat{\mathbf{x}}(i \mid j) = \mathrm{E}\left[\mathbf{x}(i) \mid \mathbf{z}(1), \cdots \mathbf{z}(j)\right].$$

This is the conditional mean, the minimum mean square error estimate. This estimate has a corresponding variance given by

$$\mathbf{P}(i \mid j) = \mathrm{E}\left[(\mathbf{x}(i) - \hat{\mathbf{x}}(i \mid j))(\mathbf{x}(i) - \hat{\mathbf{x}}(i \mid j))^T \mid \mathbf{z}(1), \cdots \mathbf{z}(j)\right]. \quad (2.3)$$

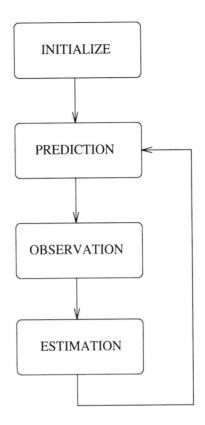

**FIGURE 2.1**
**Kalman Filter Stages**

## 2.2.2 Kalman Filter Algorithm

A great deal has been written about the Kalman filter and estimation theory in general [12], [13], [74]. An outline of the Kalman filter algorithm is presented here without derivation. Figure 2.1 summarizes its main functional stages. For a system described by Equation 2.1 and being observed according to Equation 2.2, the Kalman filter provides a recursive estimate $\hat{\mathbf{x}}(k \mid k)$ for the state $\mathbf{x}(k)$ at time $k$ given all information up to time $k$ in terms of the predicted state $\hat{\mathbf{x}}(k \mid k-1)$ and the new observation $\mathbf{z}(k)$ [41]. The *one-step-ahead* prediction, $\hat{\mathbf{x}}(k \mid k-1)$, is the estimate of the state at a time $k$ given only information up to time $(k-1)$. The Kalman filter algorithm may be summarized in two stages:

**Prediction**

$$\hat{\mathbf{x}}(k \mid k-1) = \mathbf{F}(k)\hat{\mathbf{x}}(k-1 \mid k-1) + \mathbf{B}(k)\mathbf{u}(k) \qquad (2.4)$$

$$\mathbf{P}(k \mid k-1) = \mathbf{F}(k)\mathbf{P}(k-1 \mid k-1)\mathbf{F}^T(k) + \mathbf{Q}(k). \qquad (2.5)$$

**Estimation**

$$\hat{\mathbf{x}}(k \mid k) = [\mathbf{1} - \mathbf{W}(k)\mathbf{H}(k)]\,\hat{\mathbf{x}}(k \mid k-1) + \mathbf{W}(k)\mathbf{z}(k) \qquad (2.6)$$

$$\mathbf{P}(k \mid k) = \mathbf{P}(k \mid k-1) - \mathbf{W}(k)\mathbf{S}(k)\mathbf{W}^T(k), \qquad (2.7)$$

where $\mathbf{W}(k)$ and $\mathbf{S}(k)$ known as the gain and innovation covariance matrices, respectively, are given by

$$\mathbf{W}(k) = \mathbf{P}(k \mid k-1)\mathbf{H}^T(k)\mathbf{S}^{-1}(k), \qquad (2.8)$$

$$\mathbf{S}(k) = \mathbf{H}(k)\mathbf{P}(k \mid k-1)\mathbf{H}^T(k) + \mathbf{R}(k). \qquad (2.9)$$

The matrix $\mathbf{1}$ represents the identity matrix. From Equation 2.6, the Kalman filter state estimate can be interpreted as a linear weighted sum of the state prediction and observation. The weights in this averaging process are $\{\mathbf{1} - \mathbf{W}(k)\mathbf{H}(k)\}$ associated with the prediction and $\mathbf{W}(k)$ associated with the observation. The values of the weights depend on the balance of confidence in prediction and observation as specified by the process and observation noise covariances.

## 2.3   The Information Filter

The *Information filter* is essentially a Kalman filter expressed in terms of measures of *information* about the parameters (*states*) of interest rather than direct state estimates and their associated covariances [47]. This filter has also been called the *inverse covariance* form of the Kalman filter [13], [74]. In this section, the contextual meaning of information is explained and the Information filter is derived.

### 2.3.1   Information Space

**Bayesian Theory**

The probabilistic information contained in $\mathbf{z}$ about $\mathbf{x}$ is described by the probability distribution function, $p(\mathbf{z}|\mathbf{x})$, known as the *likelihood function*. Such information is considered *objective* because it is based on observations. The likelihood function contains all the relevant information from the observation $\mathbf{z}$ required in order to make inferences about the true state $\mathbf{x}$.

This leads to the formulation of the *likelihood principle* which states that, all that is known about the unknown state is what is obtained through experimentation. Thus the likelihood function contains all the information needed to construct an estimate for $\mathbf{x}$. However, the likelihood function does not give the complete picture, if before measurement, information about the state $\mathbf{x}$ is made available exogenously. Such *a priori* information about the state is encapsulated in the prior distribution function $p(\mathbf{x})$ and is regarded as *subjective* because it is not based on any observed data. How such prior information and the likelihood information interact to provide a *posteriori* (combined prior and observed) information, is solved by *Bayes theorem* which gives the posterior conditional distribution of $\mathbf{x}$ given $\mathbf{z}$,

$$
\begin{aligned}
p(\mathbf{x}, \mathbf{z}) &= p(\mathbf{x}|\mathbf{z})p(\mathbf{z}) \\
&= p(\mathbf{z}|\mathbf{x})p(\mathbf{x}) \\
\Leftrightarrow p(\mathbf{x}|\mathbf{z}) &= \frac{p(\mathbf{z}|\mathbf{x})p(\mathbf{x})}{p(\mathbf{z})}.
\end{aligned}
\tag{2.10}
$$

where $p(\mathbf{z})$ is the marginal distribution.

In order to reduce uncertainty several measurements may be taken over time before constructing the posterior. The set of all observations up to time $k$ is defined as

$$
\mathbf{Z}^k \triangleq \{\mathbf{z}(1), \mathbf{z}(2), ..., \mathbf{z}(k)\}.
\tag{2.11}
$$

The corresponding likelihood function is given by

$$
\Lambda_k(\mathbf{x}) \triangleq p(\mathbf{Z}^k|\mathbf{x}).
\tag{2.12}
$$

This is a measure of how *"likely"* a parameter value $\mathbf{x}$ is, given that all the observations in $\mathbf{Z}^k$ are made. Thus the likelihood function serves as a measure of *evidence from data*. The posterior distribution of $\mathbf{x}$ given the set of observations $\mathbf{Z}^k$ is now computed as

$$
p(\mathbf{x}|\mathbf{Z}^k) = \frac{p(\mathbf{Z}^k|\mathbf{x})p(\mathbf{x})}{p(\mathbf{Z}^k)}.
\tag{2.13}
$$

It can also be computed recursively after each observation $\mathbf{z}(k)$ as follows:

$$
p(\mathbf{x}|\mathbf{Z}^k) = \frac{p(\mathbf{z}(k)|\mathbf{x})p(\mathbf{x}|\mathbf{Z}^{k-1})}{p(\mathbf{z}(k)|\mathbf{Z}^{k-1})}.
\tag{2.14}
$$

In this recursive form there is no need to store all the observations. Only the current observation $\mathbf{z}(k)$ at step $k$ is considered. This recursive definition has reduced memory requirements and hence it is the most commonly implemented form of Bayes theorem.

**Measures of Information**

The term information is employed in the *Fisher* sense, that is, a measure of the amount of information about a *random* state $\mathbf{x}$ present in the set of observations $\mathbf{Z}^k$, up to time $k$. The *score function*, $\mathbf{s}_k(\mathbf{x})$, is defined as the gradient of the log-likelihood function,

$$\mathbf{s}_k(\mathbf{x}) \overset{\triangle}{=} \nabla_x ln\ p(\mathbf{Z}^k, \mathbf{x}) = \frac{\nabla_x p(\mathbf{Z}^k, \mathbf{x})}{p(\mathbf{Z}^k, \mathbf{x})}. \tag{2.15}$$

By considering $\mathbf{s}_k(\mathbf{x})$ as a random variable, its mean is obtained from

$$\mathrm{E}\left[\mathbf{s}_k(\mathbf{x})\right] = \int \frac{\nabla_x p(\mathbf{Z}^k, \mathbf{x})}{p(\mathbf{Z}^k, \mathbf{x})} p(\mathbf{Z}^k, \mathbf{x}) d\mathbf{z}$$

$$= \nabla_x \int p(\mathbf{Z}^k, \mathbf{x}) d\mathbf{z} = 0.$$

The *Fisher information matrix* $\mathcal{J}(k)$ is then defined as the covariance of the score function,

$$\mathcal{J}(k) \overset{\triangle}{=} \mathrm{E}\left[\{\nabla_x ln\ p(\mathbf{Z}^k, \mathbf{x})\}\{\nabla_x ln\ p(\mathbf{Z}^k, \mathbf{x})\}^T\right]. \tag{2.16}$$

Expressing this result as the negative expectation of the Hessian of the log-likelihood gives

$$\mathcal{J}(k) = -\mathrm{E}\left[\nabla_x \nabla_x^T ln\ p(\mathbf{Z}^k, \mathbf{x})\right]. \tag{2.17}$$

For a *non-random* state $\mathbf{x}$ the expression of the Fisher information matrix becomes

$$\mathcal{J}(k) = -\mathrm{E}\left[\nabla_x \nabla_x^T ln\ p(\mathbf{Z}^k | \mathbf{x})\right]. \tag{2.18}$$

The notion of Fisher information is useful in estimation and control. It is consistent with information in the sense of the *Cramer-Rao lower bound* (**CRLB**) [13]. According to the CRLB, the mean squared error corresponding to the estimator of a parameter cannot be smaller than a certain quantity related to the likelihood function. Thus the CRLB bounds the mean squared error vector of any unbiased estimator $\hat{\mathbf{x}}(k \mid k)$ for a state vector $\mathbf{x}(k)$ modeled as random.

$$\mathrm{E}[\{\mathbf{x}(k) - \hat{\mathbf{x}}(k \mid k)\}\{\mathbf{x}(k) - \hat{\mathbf{x}}(k \mid k)\}^T | \mathbf{Z}^k] \geq \mathcal{J}^{-1}(k). \tag{2.19}$$

In this way the covariance matrix of an unbiased estimator is bounded from below. It follows from Equation 2.19 that the CRLB is the inverse of the Fisher information matrix, $\mathcal{J}(k)$. This is a very important relationship. A necessary condition for an estimator to be *consistent* in the mean square sense is that there must be an increasing amount of information (in the sense of Fisher) about the parameter in the measurements, i.e., the Fisher information has to tend to infinity as $k \to \infty$. The CRLB then converges to zero as $k \to \infty$ and thus the variance can also converge to zero. Furthermore, if an estimator's variance is equal to the CRLB, then such an estimator is called *efficient*.

Consider the expression for the Fisher information matrix in Equations 2.16 or 2.17. In the particular case where the likelihood function, $\Lambda_k(\mathbf{x})$, is *Gaussian*, it can be shown that the Fisher information matrix, $\mathcal{J}(k)$, is equal to the inverse of the covariance matrix $\mathbf{P}(k \mid k)$, that is, the CRLB is the covariance matrix. This is done by considering the probability distribution function of a *Gaussian* random vector $\mathbf{x}(k)$ whose mean and associated covariance matrix are $\hat{\mathbf{x}}(k \mid k)$ and $\mathbf{P}(k \mid k)$, respectively. In particular,

$$p(\mathbf{x}(k)|\mathbf{Z}^k) = \mathcal{N}(\mathbf{x}(k), \hat{\mathbf{x}}(k \mid k), \mathbf{P}(k \mid k))$$

$$\triangleq \frac{1}{\mathbf{A}} \exp\left\{ -\frac{[\mathbf{x}(k) - \hat{\mathbf{x}}(k \mid k)]^T \mathbf{P}^{-1}(k \mid k)[\mathbf{x}(k) - \hat{\mathbf{x}}(k \mid k)]}{2} \right\},$$

where $\mathbf{A} = \sqrt{det(2\pi\mathbf{P}(k \mid k))}$. Substituting this distribution into Equation 2.17 leads to

$$\mathcal{J}(k) = -\mathrm{E}\left[\nabla_x \nabla_x^T ln\, p(\mathbf{x}(k)|\mathbf{Z}^k)\right]$$

$$= \mathrm{E}\left[\nabla_x \nabla_x^T \left\{ \frac{[\mathbf{x}(k) - \hat{\mathbf{x}}(k \mid k)]^T \mathbf{P}^{-1}(k \mid k)[\mathbf{x}(k) - \hat{\mathbf{x}}(k \mid k)]}{2} + ln\, \mathbf{A} \right\}\right]$$

$$= \mathrm{E}\left[\nabla_x \nabla_x^T \left( \frac{[\mathbf{x}(k) - \hat{\mathbf{x}}(k \mid k)]^T \mathbf{P}^{-1}(k \mid k)[\mathbf{x}(k) - \hat{\mathbf{x}}(k \mid k)]}{2} \right)\right]$$

$$= \mathrm{E}\left[\mathbf{P}^{-1}(k \mid k)\left\{[\mathbf{x}(k) - \hat{\mathbf{x}}(k \mid k)][\mathbf{x}(k) - \hat{\mathbf{x}}(k \mid k)]^T\right\}\mathbf{P}^{-1}(k \mid k)\right]$$

$$= \mathbf{P}^{-1}(k \mid k)\mathbf{P}(k \mid k)\mathbf{P}^{-1}(k \mid k)$$

$$= \mathbf{P}^{-1}(k \mid k) \qquad\qquad (2.20)$$

$$= (CRLB)^{-1}. \qquad\qquad (2.21)$$

Thus, assuming Gaussian noise and minimum mean squared error estimation, the Fisher information matrix is equal to the inverse of the covariance matrix.

This *information matrix* is central to the filtering techniques employed in this book. Although the filter constructed from this information space is algebraically equivalent to the Kalman filter, it has been shown to have advantages over the Kalman filter in multisensor data fusion applications. These include reduced computation, algorithmic simplicity and easy initialization. In particular, these attributes make the information filter easier to decouple, decentralize and distribute. These are important filter characteristics in multisensor data fusion systems.

## 2.3.2   Information Filter Derivation

The two key information-analytic variables are the *information matrix* and *information state vector*. The information matrix has already been derived above (Section 2.3.1) as the inverse of the covariance matrix,

$$\mathbf{Y}(i \mid j) \stackrel{\triangle}{=} \mathbf{P}^{-1}(i \mid j). \tag{2.22}$$

The information state vector is a product of the inverse of the covariance matrix (information matrix) and the state estimate,

$$\begin{aligned}
\hat{\mathbf{y}}(i \mid j) &\stackrel{\triangle}{=} \mathbf{P}^{-1}(i \mid j)\hat{\mathbf{x}}(i \mid j) \\
&= \mathbf{Y}(i \mid j)\hat{\mathbf{x}}(i \mid j)
\end{aligned} \tag{2.23}$$

The variables, $\mathbf{Y}(i \mid j)$ and $\hat{\mathbf{y}}(i \mid j)$, form the basis of the information space ideas which are central to the material presented in this book.

The Information filter is derived from the Kalman filter algorithm by post-multiplying the term $\{\mathbf{1} - \mathbf{W}(k)\mathbf{H}(k)\}$ from Equation 2.6, by the term $\{\mathbf{P}(k \mid k-1)\mathbf{P}^{-1}(k \mid k-1)\}$ (i.e., post-multiplication by the identity matrix $\mathbf{1}$),

$$\begin{aligned}
\mathbf{1} - \mathbf{W}(k)\mathbf{H}(k) &= \left[\mathbf{P}(k \mid k-1) - \mathbf{W}(k)\mathbf{H}(k)\mathbf{P}(k \mid k-1)\right]\mathbf{P}^{-1}(k \mid k-1) \\
&= \left[\mathbf{P}(k \mid k-1) - \mathbf{W}(k)\mathbf{S}(k)\mathbf{S}^{-1}(k)\mathbf{H}(k)\mathbf{P}(k \mid k-1)\right] \times \\
&\quad \mathbf{P}^{-1}(k \mid k-1) \\
&= \left[\mathbf{P}(k \mid k-1) - \mathbf{W}(k)\mathbf{S}(k)\mathbf{W}^{T}(k)\right]\mathbf{P}^{-1}(k \mid k-1) \\
&= \mathbf{P}(k \mid k)\mathbf{P}^{-1}(k \mid k-1). \tag{2.24}
\end{aligned}$$

Substituting the expression of the innovation covariance $\mathbf{S}(k)$, given in Equation 2.9, into the expression of the filter gain matrix $\mathbf{W}(k)$, from Equation 2.8 gives

$$\begin{aligned}
\mathbf{W}(k) &= \mathbf{P}(k \mid k-1)\mathbf{H}^{T}(k)[\mathbf{H}(k)\mathbf{P}(k \mid k-1)\mathbf{H}^{T}(k) + \mathbf{R}(k)]^{-1} \\
\Leftrightarrow \mathbf{W}(k)&[\mathbf{H}(k)\mathbf{P}(k \mid k-1)\mathbf{H}^{T}(k) + \mathbf{R}(k)] = \mathbf{P}(k \mid k-1)\mathbf{H}^{T}(k) \\
\Leftrightarrow \mathbf{W}(k)&\mathbf{R}(k) = [\mathbf{1} - \mathbf{W}(k)\mathbf{H}(k)]\mathbf{P}(k \mid k-1)\mathbf{H}^{T}(k)
\end{aligned}$$

$$\Leftrightarrow \mathbf{W}(k) = [\mathbf{1} - \mathbf{W}(k)\mathbf{H}(k)]\mathbf{P}(k \mid k-1)\mathbf{H}^{T}(k)\mathbf{R}^{-1}(k). \tag{2.25}$$

Substituting Equation 2.24 into Equation 2.25 gives

$$\mathbf{W}(k) = \mathbf{P}(k \mid k)\mathbf{H}^{T}(k)\mathbf{R}^{-1}(k). \tag{2.26}$$

Substituting Equations 2.24 and 2.26 into Equation 2.6 and pre-multiplying through by $\mathbf{P}^{-1}(k \mid k)$ gives the update equation for the information state vector as

$$\mathbf{P}^{-1}(k \mid k)\hat{\mathbf{x}}(k \mid k) = \mathbf{P}^{-1}(k \mid k-1)\hat{\mathbf{x}}(k \mid k-1) + \mathbf{H}^{T}(k)\mathbf{R}^{-1}(k)\mathbf{z}(k),$$

or

$$\hat{y}(k \mid k) = \hat{y}(k \mid k-1) + \mathbf{H}^T(k)\mathbf{R}^{-1}(k)\mathbf{z}(k). \tag{2.27}$$

A similar expression can be found for the information matrix associated with this estimate. From Equations 2.7, 2.8 and 2.24 it follows that

$$\mathbf{P}(k \mid k) = [\mathbf{1} - \mathbf{W}(k)\mathbf{H}(k)]\,\mathbf{P}(k \mid k-1)[\mathbf{1} - \mathbf{W}(k)\mathbf{H}(k)]^T$$
$$+\mathbf{W}(k)\mathbf{R}(k)\mathbf{W}^T(k). \tag{2.28}$$

Substituting in Equations 2.24 and 2.26 gives

$$\mathbf{P}(k \mid k) = \left[\mathbf{P}(k \mid k)\mathbf{P}^{-1}(k \mid k-1)\right]\mathbf{P}(k \mid k-1)\left[\mathbf{P}(k \mid k)\mathbf{P}^{-1}(k \mid k-1)\right]^T$$
$$+ \left[\mathbf{P}(k \mid k)\mathbf{H}^T(k)\mathbf{R}^{-1}(k)\right]\mathbf{R}(k)\left[\mathbf{P}(k \mid k)\mathbf{H}^T(k)\mathbf{R}^{-1}(k)\right]^T. \tag{2.29}$$

Pre- and post-multiplying by $\mathbf{P}^{-1}(k \mid k)$ then simplifying gives the information matrix update equation as

$$\mathbf{P}^{-1}(k \mid k) = \mathbf{P}^{-1}(k \mid k-1) + \mathbf{H}^T(k)\mathbf{R}^{-1}(k)\mathbf{H}(k) \tag{2.30}$$

or

$$\mathbf{Y}(k \mid k) = \mathbf{Y}(k \mid k-1) + \mathbf{H}^T(k)\mathbf{R}^{-1}(k)\mathbf{H}(k). \tag{2.31}$$

The information state contribution $\mathbf{i}(k)$ from an observation $\mathbf{z}(k)$, and its associated information matrix $\mathbf{I}(k)$ are defined, respectively, as follows:

$$\mathbf{i}(k) \triangleq \mathbf{H}^T(k)\mathbf{R}^{-1}(k)\mathbf{z}(k), \tag{2.32}$$

$$\mathbf{I}(k) \triangleq \mathbf{H}^T(k)\mathbf{R}^{-1}(k)\mathbf{H}(k). \tag{2.33}$$

The information propagation coefficient $\mathbf{L}(k \mid k-1)$, which is independent of the observations made, is given by the expression

$$\mathbf{L}(k \mid k-1) = \mathbf{Y}(k \mid k-1)\mathbf{F}(k)\mathbf{Y}^{-1}(k-1 \mid k-1). \tag{2.34}$$

With these information quantities well defined, the linear Kalman filter can now be written in terms of the information state vector and the information matrix.

**Prediction**

$$\hat{y}(k \mid k-1) = \mathbf{L}(k \mid k-1)\hat{y}(k-1 \mid k-1) \tag{2.35}$$

$$\mathbf{Y}(k \mid k-1) = \left[\mathbf{F}(k)\mathbf{Y}^{-1}(k-1 \mid k-1)\mathbf{F}^T(k) + \mathbf{Q}(k)\right]^{-1}. \tag{2.36}$$

**Estimation**

$$\hat{y}(k \mid k) = \hat{y}(k \mid k-1) + \mathbf{i}(k) \tag{2.37}$$

$$\mathbf{Y}(k \mid k) = \mathbf{Y}(k \mid k-1) + \mathbf{I}(k). \qquad (2.38)$$

This is the information form of the Kalman filter [46], [87], [48]. Despite its potential applications, it is not widely used and it is thinly covered in literature. Barshalom [13] and Maybeck [74] briefly discuss the idea of information estimation, but do not explicitly derive the algorithm in terms of information as done above, nor do they use it as a principal filtering method.

### 2.3.3  Filter Characteristics

By comparing the implementation requirements and performance of the Kalman and Information filters, a number of attractive features of the latter are identified:

- The information estimation Equations 2.37 and 2.38 are computationally simpler than the state estimation Equations 2.6 and 2.7. This can be exploited in partitioning these equations for decentralized multisensor estimation.

- Although the information prediction Equations 2.35 and 2.36 are more complex than Equations 2.4 and 2.5, prediction depends on a propagation coefficient which is independent of the observations. It is thus again easy to decouple and decentralize.

- There are no gain or innovation covariance matrices and the maximum dimension of a matrix to be inverted is the state dimension. In multisensor systems the state dimension is generally smaller than the observation dimension, hence it is preferable to employ the Information filter and invert smaller information matrices than use the Kalman filter and invert larger innovation covariance matrices.

- Initializing the Information filter is much easier than for the Kalman filter. This is because information estimates (matrix and state) are easily initialized to zero information. However, in order to implement the Information filter, a start-up procedure is required where the information matrix is set with small non-zero diagonal elements to make it invertible.

These characteristics are useful in the development of decentralized data fusion and control systems. Consequently, this book employs information space estimation as the principal filtering technique.

### 2.3.4  An Example of Linear Estimation

To compare the Kalman and the Information filter and illustrate the issues discussed above, the following example of a linear estimation problem is considered. Consider two targets moving with two different but constant velocities, $v_1$ and $v_2$. The state vector describing their true positions and velocities can be represented as follows:

$$\mathbf{x}(k) = \begin{bmatrix} x_1(k) \\ x_2(k) \\ \dot{x}_1(k) \\ \dot{x}_2(k) \end{bmatrix} = \begin{bmatrix} v_1 k \\ v_2 k \\ v_1 \\ v_2 \end{bmatrix}. \tag{2.39}$$

The objective is to estimate the entire state vector $\mathbf{x}(k)$ in Equation 2.39 after obtaining observations of the two target positions, $x_1(k)$ and $x_2(k)$.

The discrete time state equation with sampling interval $\Delta T$ is given by

$$\mathbf{x}(k) = \mathbf{F}(k)\mathbf{x}(k-1) + \mathbf{w}(k-1), \tag{2.40}$$

where $\mathbf{F}(k)$ is the state transition matrix. This matrix is obtained by the Series method as follows:

$$\mathbf{F}(k) = e^{\mathbf{A}\Delta T} \approx \mathbf{1} + \Delta T \mathbf{A}$$

$$= \begin{bmatrix} 1 & 0 & \Delta T & 0 \\ 0 & 1 & 0 & \Delta T \\ 0 & 0 & 1 & 0 \\ 0 & 0 & 0 & 1 \end{bmatrix},$$

where $\mathbf{1}$ is an identity matrix and A is given by

$$\mathbf{A} = \begin{bmatrix} 0 & 0 & 1 & 0 \\ 0 & 0 & 0 & 1 \\ 0 & 0 & 0 & 0 \\ 0 & 0 & 0 & 0 \end{bmatrix}.$$

Since only linear measurements of the two target positions are taken, the observation matrix is given by

$$\mathbf{H}(k) = \begin{bmatrix} 1 & 0 & 0 & 0 \\ 0 & 1 & 0 & 0 \end{bmatrix}.$$

In order to complete the construction of models, the measurement error covariance matrix $\mathbf{R}(k)$ and the process noise $\mathbf{Q}(k)$ are then obtained as follows:

$$\mathbf{R}(k) = \begin{bmatrix} \sigma^2_{meas\_noise} & 0 \\ 0 & \sigma^2_{meas\_noise} \end{bmatrix},$$

$$\mathbf{Q}(k) = \begin{bmatrix} \sigma^2_{pos\_noise} & 0 & 0 & 0 \\ 0 & \sigma^2_{pos\_noise} & 0 & 0 \\ 0 & 0 & \sigma^2_{vel\_noise} & 0 \\ 0 & 0 & 0 & \sigma^2_{vel\_noise} \end{bmatrix}$$

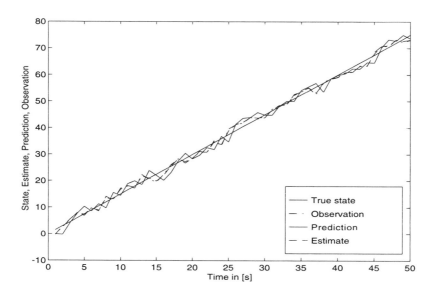

**FIGURE 2.2**
**Performance of the Kalman and Information Filters**

The terms $\sigma_{pos\_noise}$ and $\sigma_{vel\_noise}$ represent the system modeling errors in target position and velocity, respectively, while $\sigma_{meas\_noise}$ represents the error in measuring a target position. The corresponding process and measurement noise vectors are defined and generated as follows:

$$\mathbf{w}(k) = \begin{bmatrix} |\text{rand num}| \leq 2\sigma_{pos\_noise} \\ |\text{rand num}| \leq 2\sigma_{pos\_noise} \\ |\text{rand num}| \leq 2\sigma_{vel\_noise} \\ |\text{rand num}| \leq 2\sigma_{vel\_noise} \end{bmatrix},$$

$$\mathbf{v}(k) = \begin{bmatrix} |\text{rand num}| \leq 2\sigma_{meas\_noise} \\ |\text{rand num}| \leq 2\sigma_{meas\_noise} \end{bmatrix}.$$

These system modeling matrices and vectors are then used in the algorithms of the Kalman and Information filters to carry out estimation. In both cases the simulations are implemented using the same models with process and observation noises generated by the same random generators. The results are discussed in the next section.

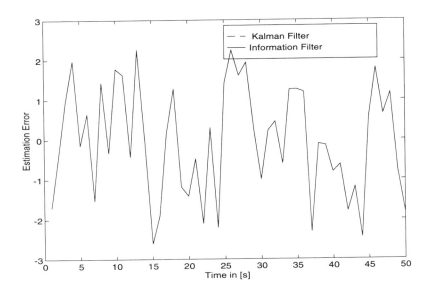

**FIGURE 2.3**
**Estimation Error for the Kalman and Information Filters**

## 2.3.5 Comparison of the Kalman and Information Filters

The Kalman and Information filters are compared by simulating the constant velocity system described above. In order to study and compare the performance of the filters, estimation of the same state is considered for the two filters; the position of the first target, $x_1(k)$. Figure 2.2 compares the target's true position, predicted position, estimated position and observed position for both the Kalman and Information filters.

The curves depicting the same variables are identical and indistinguishable for the two filters. They lie on top of each other. This illustrates the algebraic equivalence of the two filters. From Figure 2.2 it can be observed, for both filters, that the state estimate is always well placed between the observation and state prediction. This means that there is balanced confidence in observations and predictions. Since as the time $k$ goes to infinity, the process noise variance $\mathbf{Q}(k)$ governs the confidence in predictions and the observation noise variance $\mathbf{R}(k)$ governs the confidence in observations, the results are an indication that the noise variances were well chosen. Figure 2.3 shows the state estimation errors for both the Kalman and Information filters while the innovations are similarly shown in Figure 2.4. The estimation errors and the innovations are identical for the two filters, again demonstrating the algebraic equivalence of the two filters. By inspection and computing the sequence mean, the innovations are shown to be zero

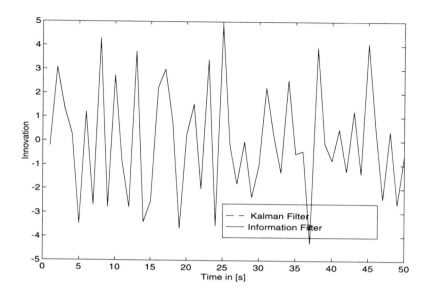

**FIGURE 2.4**
**Innovations for the Kalman and Information Filters**

mean with variance $\mathbf{S}(k)$. Practically, it means the noise level in the filter is of the same order as the true system noise. There is no visible correlation of the innovations sequences. This implies that there are no significant higher-order unmodeled dynamics nor excessive observation noise to process noise ratio. The innovations also satisfy the 95% confidence rule. This implies that the filters are consistent and well-matched. Since the curves in Figures 2.2, 2.3 and 2.4 look indistinguishable for the two filters, it is prudent to plot parameter differences between the filters to confirm the algebraic equivalence. Figure 2.5 shows the difference between the state estimates for the filters. The difference is very small (*lies within* $10^{-13}\%$) and hence attributable to numerical and computational errors such as truncation and rounding off errors. Thus the Kalman and Information filters are demonstrably equivalent. This confirms the algebraic equivalence which is mathematically proven and established in the derivation of the Information filter from the Kalman filter.

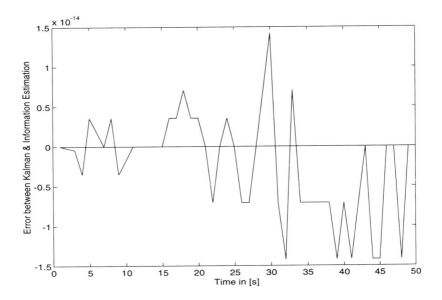

**FIGURE 2.5**
**The Difference between Kalman and Information Filters' State Estimates**

## 2.4 The Extended Kalman Filter (EKF)

In almost all real data fusion problems the state or environment of interest does not evolve linearly. Consequently simple linear models will not be adequate to describe the system. Furthermore, the sensor observations may not depend linearly on the states that describe the environment. A popular approach to solve nonlinear estimation problems has been to use the extended Kalman filter (**EKF**) [12], [24]. This is a *linear* estimator for a nonlinear system obtained by *linearization* of the nonlinear state and observations equations. For any nonlinear system the EKF is the best linear, unbiased, estimator with respect to minimum mean squared error criteria.

The EKF is conceptually simple and its derivation follows from arguments of linearization and the Kalman filter algorithm. The difficulty arises in implementation. It can be made to work well, but may perform badly or even become unstable with diverging estimates. This is most often due to lack of careful modeling of sensors and environment. Failure to understand the limitations of the algorithm exacerbates the problem.

### 2.4.1  Nonlinear State Space

The system of interest is described by a nonlinear discrete-time state transition equation in the form

$$\mathbf{x}(k) = \mathbf{f}\left(\mathbf{x}(k-1), \mathbf{u}(k-1), (k-1)\right) + \mathbf{w}(k), \qquad (2.41)$$

where $\mathbf{x}(k-1)$ is the state vector and $\mathbf{u}(k-1)$ is a known input vector, both at time $(k-1)$. The vectors $\mathbf{x}(k)$ and $\mathbf{w}(k)$ represent the state vector and some additive process noise vector, respectively, both at time-step $k$. The nonlinear function $\mathbf{f}(\cdot, \cdot, k)$ is the nonlinear state transition function mapping the previous state and current control input to the current state. It is assumed that observations of the state of this system are made according to a nonlinear observation equation of the form

$$\mathbf{z}(k) = \mathbf{h}\left(\mathbf{x}(k), k\right) + \mathbf{v}(k), \qquad (2.42)$$

where $\mathbf{z}(k)$ is the observation made at time $k$, $\mathbf{x}(k)$ is the state at time $k$, $\mathbf{v}(k)$ is some additive observation noise, and $\mathbf{h}(\cdot, k)$ is a nonlinear observation model mapping current state to observations.

It is assumed that both noise vectors $\mathbf{v}(k)$ and $\mathbf{w}(k)$ are linearly additive Gaussian, temporally uncorrelated with zero mean, which means

$$E[\mathbf{w}(k)] = E[\mathbf{v}(k)] = \mathbf{0}, \qquad \forall k,$$

with the corresponding covariances being given by

$$E[\mathbf{w}(i)\mathbf{w}^T(j)] = \delta_{ij}\mathbf{Q}(i), \quad E[\mathbf{v}(i)\mathbf{v}^T(j)] = \delta_{ij}\mathbf{R}(i).$$

It is also assumed that the process and observation noises are uncorrelated, i.e.,

$$E[\mathbf{w}(i)\mathbf{v}^T(j)] = \mathbf{0}, \qquad \forall i, j.$$

### 2.4.2  EKF Derivation

The derivation of the EKF follows from that of the linear Kalman filter, by linearizing state and observation models using Taylor's series expansion [13], [74].

### State Prediction

It is assumed that there exists an estimate at time $(k-1)$ which is approximately equal to the conditional mean,

$$\hat{\mathbf{x}}(k-1 \mid k-1) \approx E[\mathbf{x}(k-1) \mid Z^{k-1}]. \qquad (2.43)$$

The objective is to find a prediction $\hat{\mathbf{x}}(k \mid k-1)$ for the state at the next time $k$ based only on the information available up to time $(k-1)$. Expanding

Equation 2.41 as a Taylor series about the estimate $\hat{\mathbf{x}}(k-1\mid k-1)$, the following expression is obtained.

$$
\begin{aligned}
\mathbf{x}(k) = {} & \mathbf{f}(\hat{\mathbf{x}}(k-1\mid k-1), \mathbf{u}(k-1), (k-1)) \\
& + \nabla \mathbf{f}_x(k)\left[\mathbf{x}(k-1) - \hat{\mathbf{x}}(k-1\mid k-1)\right] \\
& + O\left(\left[\mathbf{x}(k-1) - \hat{\mathbf{x}}(k-1\mid k-1)\right]^2\right) + \mathbf{w}(k)
\end{aligned}
\tag{2.44}
$$

where $\nabla \mathbf{f}_x(k)$ is the Jacobian of $\mathbf{f}$ evaluated at $\mathbf{x}(k-1) = \hat{\mathbf{x}}(k-1\mid k-1)$.

Truncating Equation 2.44 at first order, and taking expectations conditioned on the first $(k-1)$ observations gives an equation for the state prediction.

$$
\begin{aligned}
\hat{\mathbf{x}}(k\mid k-1) &= \mathrm{E}\left[\mathbf{x}(k)\mid Z^{k-1}\right] \\
&\approx \mathrm{E}\left[\mathbf{f}\left(\hat{\mathbf{x}}(k-1\mid k-1) + \mathbf{A} + \mathbf{w}(k)\mid Z^{k-1}, \mathbf{u}(k-1), (k-1)\right)\right] \\
&\quad (\text{where} \quad \mathbf{A} = \nabla \mathbf{f}_x(k)\left[\mathbf{x}(k-1) - \hat{\mathbf{x}}(k-1\mid k-1)\right]) \\
&= \mathbf{f}(\hat{\mathbf{x}}(k-1\mid k-1), \mathbf{u}(k-1), (k-1)).
\end{aligned}
\tag{2.45}
$$

This follows from the assumption that the estimate $\hat{\mathbf{x}}(k-1\mid k-1)$ is approximately equal to the conditional mean (Equation 2.43) and that the process noise $\mathbf{w}(k)$ has zero mean. The state estimate error at a time $i$ given all observations up to time $j$ is defined as

$$
\tilde{\mathbf{x}}(i\mid j) \triangleq \mathbf{x}(i) - \hat{\mathbf{x}}(i\mid j),
\tag{2.46}
$$

and the state covariance is defined as the outer product of this error with itself conditioned on the observations made

$$
\mathbf{P}(i\mid j) \triangleq \mathrm{E}\left[\tilde{\mathbf{x}}(i\mid j)\tilde{\mathbf{x}}^T(i\mid j)\mid Z^j\right].
\tag{2.47}
$$

In particular, the prediction error $\tilde{\mathbf{x}}(k\mid k-1)$ can be found by subtracting the true state $\mathbf{x}(k)$ given in Equation 2.44 from the prediction given in Equation 2.45

$$
\begin{aligned}
& \tilde{\mathbf{x}}(k\mid k-1) \\
&= \mathbf{x}(k) - \hat{\mathbf{x}}(k\mid k-1) \\
&= \mathbf{f}(\hat{\mathbf{x}}(k-1\mid k-1), \mathbf{u}(k), k) + \nabla \mathbf{f}_x(k)\left[\mathbf{x}(k-1) - \hat{\mathbf{x}}(k-1\mid k-1)\right] \\
&\quad + O\left(\left[\mathbf{x}(k-1) - \hat{\mathbf{x}}(k-1\mid k-1)\right]^2\right) + \mathbf{w}(k) \\
&\quad - \mathbf{f}(\hat{\mathbf{x}}(k-1\mid k-1), \mathbf{u}(k), k) \\
&\approx \nabla \mathbf{f}_x(k)\left[\mathbf{x}(k-1) - \hat{\mathbf{x}}(k-1\mid k-1)\right] + \mathbf{w}(k) \\
&= \nabla \mathbf{f}_x(k)\tilde{\mathbf{x}}(k-1\mid k-1) + \mathbf{w}(k).
\end{aligned}
\tag{2.48}
$$

The prediction is unbiased when the previous estimate is unbiased and the condition that the noise sequences are zero mean and white hold.

Taking expectations conditioned on the observations made up to time $(k-1)$ of the outer product of the prediction error gives an expression for the prediction covariance in terms of the covariance of the previous estimate.

$$
\begin{aligned}
\mathbf{P}(k \mid k-1) \; &\triangleq \; \mathrm{E}\left[\tilde{\mathbf{x}}(k \mid k-1)\tilde{\mathbf{x}}^T(k \mid k-1) \mid Z^{k-1}\right] \\
&\approx \mathrm{E}\left[(\nabla \mathbf{f}_x(k)\tilde{\mathbf{x}}(k-1 \mid k-1) + \mathbf{w}(k))\,\mathbf{A} \mid Z^{k-1}\right] \\
&\quad \left(where \quad \mathbf{A} = (\nabla \mathbf{f}_x(k)\tilde{\mathbf{x}}(k-1 \mid k-1) + \mathbf{w}(k))^T\right) \\
&= \nabla \mathbf{f}_x(k)\mathrm{E}\left[\tilde{\mathbf{x}}(k-1 \mid k-1)\tilde{\mathbf{x}}^T(k-1 \mid k-1) \mid Z^{k-1}\right]\nabla \mathbf{f}_x^T(k) \\
&\quad + \mathrm{E}\left[\mathbf{w}(k)\mathbf{w}^T(k)\right] \\
&= \nabla \mathbf{f}_x(k)\mathbf{P}(k-1 \mid k-1)\nabla \mathbf{f}_x^T(k) + \mathbf{Q}(k).
\end{aligned}
\tag{2.49}
$$

The last two lines follow from the fact that the estimate and true state at time $(k-1)$ are statistically dependent only on the noise terms $\mathbf{v}(j)$ and $\mathbf{w}(j)$, $j \leq (k-1)$. Hence, by assumption, they are uncorrelated with the *current* process noise $\mathbf{w}(k)$.

**Observation Prediction and Innovation**

The next objective is to obtain a predicted observation and its corresponding innovation to be used in updating the predicted state. This is achieved by expanding Equation 2.42, describing the observations made, as a Taylor series about the state prediction $\hat{\mathbf{x}}(k \mid k-1)$.

$$
\begin{aligned}
\mathbf{z}(k) &= \mathbf{h}\left(\mathbf{x}(k)\right) + \mathbf{v}(k) \\
&= \mathbf{h}\left(\hat{\mathbf{x}}(k \mid k-1)\right) + \nabla \mathbf{h}_x(k)\left[\hat{\mathbf{x}}(k \mid k-1) - \mathbf{x}(k)\right] + \\
&\quad O\left([\hat{\mathbf{x}}(k \mid k-1) - \mathbf{x}(k)]^2\right) + \mathbf{v}(k)
\end{aligned}
\tag{2.50}
$$

where $\nabla \mathbf{h}_x(k)$ is the Jacobian of $\mathbf{h}$ evaluated at $\mathbf{x}(k) = \hat{\mathbf{x}}(k \mid k-1)$. Again, ignoring second and higher order terms and taking expectations conditioned on the first $(k-1)$ observations gives an equation for the predicted observation.

$$
\begin{aligned}
&\hat{\mathbf{z}}(k \mid k-1) \\
&\triangleq \; \mathrm{E}\left[\mathbf{z}(k) \mid Z^{k-1}\right] \\
&\approx \mathrm{E}\left[\mathbf{h}\left(\hat{\mathbf{x}}(k \mid k-1)\right) + \nabla \mathbf{h}_x(k)\left[\hat{\mathbf{x}}(k \mid k-1) - \mathbf{x}(k)\right] + \mathbf{v}(k) \mid Z^{k-1}\right] \\
&= \mathbf{h}\left(\hat{\mathbf{x}}(k \mid k-1)\right),
\end{aligned}
\tag{2.51}
$$

where the last two lines follow from the fact that the state prediction error and the observation noise both have zero mean. After taking an observation $\mathbf{z}(k)$, the innovation can be found by subtracting the predicted observation as

$$
\nu(k) = \mathbf{z}(k) - \mathbf{h}\left(\hat{\mathbf{x}}(k \mid k-1)\right).
\tag{2.52}
$$

The innovation covariance can now be found from the mean squared error in the predicted observation. The error in the predicted observation can be approximated by subtracting this prediction from the Taylors series expansion of the observation in Equation 2.50 as

$$
\begin{aligned}
\tilde{\mathbf{z}}(k \mid k-1) &\triangleq \mathbf{z}(k) - \hat{\mathbf{z}}(k \mid k-1) \\
&= \mathbf{h}\left(\hat{\mathbf{x}}(k \mid k-1)\right) + \nabla\mathbf{h}_x(k)\left[\hat{\mathbf{x}}(k \mid k-1) - \mathbf{x}(k)\right] \\
&\quad + O\left(\left[\hat{\mathbf{x}}(k \mid k-1) - \mathbf{x}(k)\right]^2\right) + \mathbf{v}(k) \\
&\quad - \mathbf{h}\left(\hat{\mathbf{x}}(k \mid k-1)\right) \\
&\approx \nabla\mathbf{h}_x(k)\left[\hat{\mathbf{x}}(k \mid k-1) - \mathbf{x}(k)\right] + \mathbf{v}(k). \quad (2.53)
\end{aligned}
$$

A clear distinction is made between the 'estimated' observation error $\tilde{\mathbf{z}}(k \mid k-1)$ and the measured observation error, the innovation, $\nu(k)$. Squaring the expression for the estimated observation error and taking expectation conditions on the first $(k-1)$ measurements gives an equation for the innovation covariance.

$$
\begin{aligned}
\mathbf{S}(k) &= \mathrm{E}\left[\tilde{\mathbf{z}}(k \mid k-1)\tilde{\mathbf{z}}^T(k \mid k-1)\right] \\
&= \mathrm{E}\left[\mathbf{A}\mathbf{A}^T\right] \\
&\quad (where \quad \mathbf{A} = (\nabla\mathbf{h}_x(k)\left[\hat{\mathbf{x}}(k \mid k-1) - \mathbf{x}(k)\right] + \mathbf{v}(k))) \\
&= \nabla\mathbf{h}_x(k)\mathbf{P}(k \mid k-1)\nabla\mathbf{h}_x^T(k) + \mathbf{R}(k). \quad (2.54)
\end{aligned}
$$

This follows from the fact that the state prediction is dependent only on the noise terms $\mathbf{v}(j)$ and $\mathbf{w}(j)$, $j \leq (k-1)$. Consequently, by assumption, it is statistically uncorrelated with the current observation noise $\mathbf{v}(k)$.

**Update Equations**

Equipped with the prediction and innovation equations, a *linearized* estimator can be proposed. It gives a state estimate vector $\hat{\mathbf{x}}(k \mid k)$ for the state vector $\mathbf{x}(k)$ which is described by the nonlinear state transition of Equation 2.41 and which is being observed according to the nonlinear observation Equation 2.42. It is assumed that a prediction $\hat{\mathbf{x}}(k \mid k-1)$ for the state at time $k$ has been made on the basis of the first $(k-1)$ observations $Z^{k-1}$ according to Equation 2.45. The current observation is $\mathbf{z}(k)$. The estimator is assumed to be in the form of a linear *unbiased* average of the prediction and innovation so that,

$$
\hat{\mathbf{x}}(k \mid k) = \hat{\mathbf{x}}(k \mid k-1) + \mathbf{W}(k)\left[\mathbf{z}(k) - \mathbf{h}(\hat{\mathbf{x}}(k \mid k-1))\right]. \quad (2.55)
$$

The gain matrix $\mathbf{W}(k)$ is chosen so as to minimize conditional mean squared estimation error. This error is equal to the trace of the estimate covariance $\mathbf{P}(k \mid k)$ which itself is simply the expected value of the state error $\tilde{\mathbf{x}}(k \mid k)$ squared.

From Equation 2.55 and the approximate observation error given in Equation 2.53, the state error becomes

$$\tilde{\mathbf{x}}(k \mid k)$$
$$= \hat{\mathbf{x}}(k \mid k) - \mathbf{x}(k)$$
$$= [\hat{\mathbf{x}}(k \mid k-1) - \mathbf{x}(k)] + \mathbf{W}(k)[\mathbf{h}(\mathbf{x}(k)) - \mathbf{h}(\hat{\mathbf{x}}(k \mid k-1))] + \mathbf{W}(k)\mathbf{v}(k)$$
$$\approx [\hat{\mathbf{x}}(k \mid k-1) - \mathbf{x}(k)] - \mathbf{W}(k)\nabla \mathbf{h}_x(k)[\hat{\mathbf{x}}(k \mid k) - \mathbf{x}(k)] + \mathbf{W}(k)\mathbf{v}(k)$$
$$= [\mathbf{1} - \mathbf{W}(k)\nabla \mathbf{h}_x(k)]\,\tilde{\mathbf{x}}(k \mid k-1) + \mathbf{W}(k)\mathbf{v}(k). \tag{2.56}$$

The estimate is unbiased when the prediction is unbiased and the condition that the noise sequences are zero mean and white hold.

Taking the expectation condition on the observations made up to time $k$ of the outer product of the state error gives an expression for the state covariance in terms of the prediction covariance.

$$\mathbf{P}(k \mid k) \triangleq \mathrm{E}\left[\tilde{\mathbf{x}}(k \mid k)\tilde{\mathbf{x}}^T(k \mid k) \mid Z^k\right]$$
$$\approx [\mathbf{1} - \mathbf{W}(k)\nabla \mathbf{h}_x(k)]\,\mathrm{E}\left[\tilde{\mathbf{x}}(k \mid k-1)\tilde{\mathbf{x}}^T(k \mid k-1) \mid Z^{k-1}\right] \times$$
$$[\mathbf{1} - \mathbf{W}(k)\nabla \mathbf{h}_x(k)]^T + \mathbf{W}(k)\mathrm{E}\left[\mathbf{v}(k)\mathbf{v}^T(k)\right]\mathbf{W}^T(k)$$
$$\approx [\mathbf{1} - \mathbf{W}(k)\nabla \mathbf{h}_x(k)]\,\mathbf{P}(k \mid k-1)[\mathbf{I} - \mathbf{W}(k)\nabla \mathbf{h}_x(k)]^T +$$
$$\mathbf{W}(k)\mathbf{R}(k)\mathbf{W}^T(k). \tag{2.57}$$

The gain matrix $\mathbf{W}(k)$ is now chosen to minimize the mean squared estimation error $L(k)$ which is defined as

$$L(k) = \mathrm{E}[\tilde{\mathbf{x}}^T(k \mid k)\tilde{\mathbf{x}}(k \mid k)] = \mathrm{trace}[\mathbf{P}(k \mid k)]. \tag{2.58}$$

Minimization of this error calls for

$$\frac{\partial L}{\partial \mathbf{W}(k)} = -2(\mathbf{1} - \mathbf{W}(k)\nabla \mathbf{h}_x(k))\mathbf{P}(k \mid k-1)\nabla \mathbf{h}_x^T(k) + 2\mathbf{W}(k)\mathbf{R}(k) = \mathbf{0}, \tag{2.59}$$

which on simplification and rearrangement provides an expression for the gain matrix as

$$\mathbf{W}(k) = \mathbf{P}(k \mid k-1)\nabla \mathbf{h}_x^T(k)\left[\nabla \mathbf{h}_x(k)\mathbf{P}(k \mid k-1)\nabla \mathbf{h}_x^T(k) + \mathbf{R}(k)\right]^{-1}$$
$$= \mathbf{P}(k \mid k-1)\nabla \mathbf{h}_x^T(k)\mathbf{S}^{-1}(k). \tag{2.60}$$

With this gain matrix, Equation 2.55 becomes the best (*minimum mean squared error*) linear unbiased estimator for the state $\mathbf{x}(k)$ under the stated conditions. This completes the derivation of the extended Kalman filter.

## 2.4.3   Summary of the EKF Algorithm

**Prediction**

$$\hat{\mathbf{x}}(k \mid k-1) = \mathbf{f}\left(\hat{\mathbf{x}}(k-1 \mid k-1), \mathbf{u}(k-1), (k-1)\right) \tag{2.61}$$

$$\mathbf{P}(k \mid k-1) = \nabla \mathbf{f}_x(k)\mathbf{P}(k-1 \mid k-1)\nabla \mathbf{f}_x{}^T(k) + \mathbf{Q}(k-1). \qquad (2.62)$$

**Estimation**

$$\hat{\mathbf{x}}(k \mid k) = \hat{\mathbf{x}}(k \mid k-1) + \mathbf{W}(k)\left[\mathbf{z}(k) - \mathbf{h}(\hat{\mathbf{x}}(k \mid k-1))\right] \qquad (2.63)$$

$$\mathbf{P}(k \mid k) = \mathbf{P}(k \mid k-1) - \mathbf{W}(k)\mathbf{S}(k)\mathbf{W}^T(k). \qquad (2.64)$$

The gain and innovation covariance matrices are given by

$$\mathbf{W}(k) = \mathbf{P}(k \mid k-1)\nabla \mathbf{h}_x{}^T(k)\mathbf{S}^{-1}(k) \qquad (2.65)$$

$$\mathbf{S}(k) = \nabla \mathbf{h}_x(k)\mathbf{P}(k \mid k-1)\nabla \mathbf{h}_x{}^T(k) + \mathbf{R}(k). \qquad (2.66)$$

The Jacobians $\nabla \mathbf{f}_x(k)$ and $\nabla \mathbf{h}_x(k)$ are typically not constant, being functions of both the state and time-step. It is clearly evident that the EKF is very similar to the Kalman filter algorithm, with the substitutions $\mathbf{F} \rightarrow \nabla \mathbf{f}_x(k)$ and $\mathbf{H} \rightarrow \nabla \mathbf{h}_x(k)$ being made in the equations for the variance and gain propagation.

It is prudent to note a number of problematic issues specific to the EKF. Unlike the linear filter, the covariances and gain matrix must be computed on-line as estimates and predictions are made available, and will not, in general, tend to constant values. This significantly increases the amount of computation which must be performed on-line by the algorithm. Also, if the nominal (predicted) trajectory is too far away from the true trajectory, then the true covariance will be much larger than the estimated covariance and the filter will become poorly matched. This might lead to severe filter instabilities. Last, the EKF employs a linearized model which must be computed from an approximate knowledge of the state. Unlike the linear algorithm, this means that the filter must be accurately initialized at the start of operation to ensure that the linearized models obtained are valid. All these issues must be taken into account in order to achieve acceptable performance for the EKF.

## 2.5 The Extended Information Filter (EIF)

### 2.5.1 Nonlinear Information Space

In this section the linear Information filter is extended to a linearized estimation algorithm for nonlinear systems. The general approach is to apply the principles of the EKF and those of the linear Information filter in order to construct a new estimation method for nonlinear systems. This generates a filter that predicts and estimates information about *nonlinear* state parameters given *nonlinear* observations and *nonlinear* system dynamics.

All the computation and tracking is in information space. The new filter will be termed the extended Information filter (**EIF**) and is the major novel contribution of this chapter. In addition to providing a solution to the nonlinear estimation problem, the EIF also has all the advantages of the Information filter and resolves some of the problems associated with the EKF.

In particular, information space allows easy initialization of filter matrices and vectors. Given the importance of accurate initialization when using linearized models, this is an extremely desirable characteristic. As discussed before, a major drawback of the EKF is excessive computational burden. Carrying out the prediction and estimation processes in terms of information will significantly reduce this load by simplifying the prediction and estimation equations. These equations are then easily partitioned and decentralized. It is proposed that estimation for nonlinear systems, in particular multisensor systems, is best carried out using information variables rather than state variables.

## 2.5.2    EIF Derivation

The derivation of the extended Information filter uses principles from both the derivations of the Information filter and the EKF. The EIF cannot be extrapolated from these two filters in an obvious manner. This is because in the nonlinear case, the function operator **h** *cannot* be separated from $\mathbf{x}(k)$ in the nonlinear observation equation

$$\mathbf{z}(k) = \mathbf{h}\left(\mathbf{x}(k), k\right) + \mathbf{v}(k), \tag{2.67}$$

and yet the derivation of the Information filter *depends* on this separation, which is possible in the linear observation equation.

The derivation of the EIF proceeds by considering equations from the derivation of the EKF algorithm. Post-multiplying $\{\mathbf{1} - \mathbf{W}(k)\nabla\mathbf{h}_x(k)\}$ from Equation 2.56 by the term $\{\mathbf{P}(k \mid k - 1)\mathbf{P}^{-1}(k \mid k - 1)\}$, i.e., post-multiplication by the identity matrix $\mathbf{1}$ leads to

$$
\begin{aligned}
\mathbf{1} - \mathbf{W}(k)\nabla\mathbf{h}_x(k) &= [\mathbf{P}(k \mid k - 1) - \mathbf{W}(k)\nabla\mathbf{h}_x(k)\mathbf{P}(k \mid k - 1)] \times \\
&\quad \mathbf{P}^{-1}(k \mid k - 1) \\
&= [\mathbf{P}(k \mid k - 1) - \mathbf{W}(k)\{\mathbf{S}(k)\mathbf{S}^{-1}(k)\}\nabla\mathbf{h}_x(k) \times \\
&\quad \mathbf{P}(k \mid k - 1)]\mathbf{P}^{-1}(k \mid k - 1) \\
&= [\mathbf{P}(k \mid k - 1) - \mathbf{W}(k)\mathbf{S}(k)\mathbf{W}^T(k)]\mathbf{P}^{-1}(k \mid k - 1)
\end{aligned}
$$
$$\Leftrightarrow \mathbf{1} - \mathbf{W}(k)\nabla\mathbf{h}_x(k) = \mathbf{P}(k \mid k)\mathbf{P}^{-1}(k \mid k - 1). \tag{2.68}$$

Substituting the expression of the EKF innovation covariance matrix from Equation 2.66 in the EKF gain matrix given in Equation 2.65 produces

$$\mathbf{W}(k) = \mathbf{P}(k \mid k - 1)\nabla\mathbf{h}_x{}^T(k)[\nabla\mathbf{h}_x(k)\mathbf{P}(k \mid k - 1)\nabla\mathbf{h}_x{}^T(k) + \mathbf{R}(k)]^{-1}$$

$$\Leftrightarrow \mathbf{W}(k)[\nabla \mathbf{h}_x(k)\mathbf{P}(k \mid k-1)\nabla \mathbf{h}_x{}^T(k) + \mathbf{R}(k)] = \mathbf{P}(k \mid k-1)\nabla \mathbf{h}_x{}^T(k)$$
$$\Leftrightarrow \mathbf{W}(k)\mathbf{R}(k) = [\mathbf{1} - \mathbf{W}(k)\nabla \mathbf{h}_x(k)]\mathbf{P}(k \mid k-1)\nabla \mathbf{h}_x{}^T(k).$$

Now, substituting Equation 2.68 into this expression gives

$$\mathbf{W}(k) = \mathbf{P}(k \mid k)\nabla \mathbf{h}_x{}^T(k)\mathbf{R}^{-1}(k). \tag{2.69}$$

In order to use Equations 2.68 and 2.69 to derive the EIF, the EKF state estimation Equation 2.63 must be expressed in the same form as that for the conventional Kalman filter. This is done by adding and subtracting the term $\mathbf{W}(k)\nabla \mathbf{h}_x(k)\hat{\mathbf{x}}(k \mid k-1)$ to the left-hand side of Equation 2.63 (addition of zero):

$$\hat{\mathbf{x}}(k \mid k)$$
$$= \hat{\mathbf{x}}(k \mid k-1) + \mathbf{W}(k)\left[\mathbf{z}(k) - \mathbf{h}(\hat{\mathbf{x}}(k \mid k-1))\right]$$
$$= \hat{\mathbf{x}}(k \mid k-1) - \mathbf{W}(k)\nabla \mathbf{h}_x(k)\hat{\mathbf{x}}(k \mid k-1) + \mathbf{W}(k)\left[\mathbf{z}(k) - \mathbf{h}(\hat{\mathbf{x}}(k \mid k-1))\right]$$
$$+ \mathbf{W}(k)\nabla \mathbf{h}_x(k)\hat{\mathbf{x}}(k \mid k-1)$$
$$= [\mathbf{1} - \mathbf{W}(k)\nabla \mathbf{h}_x(k)]\hat{\mathbf{x}}(k \mid k-1) + \mathbf{W}(k)\left[\nu(k) + \nabla \mathbf{h}_x(k)\hat{\mathbf{x}}(k \mid k-1)\right]$$
$$= [\mathbf{1} - \mathbf{W}(k)\nabla \mathbf{h}_x(k)]\hat{\mathbf{x}}(k \mid k-1) + \mathbf{W}(k)\mathbf{z}'(k), \tag{2.70}$$

where $\mathbf{z}'(k)$ is the *'equivalent' linearized* observation vector,

$$\mathbf{z}'(k) = \nu(k) + \nabla \mathbf{h}_x(k)\hat{\mathbf{x}}(k \mid k-1), \tag{2.71}$$

and the innovation vector is given by

$$\nu(k) = \mathbf{z}(k) - \mathbf{h}(\hat{\mathbf{x}}(k \mid k-1)). \tag{2.72}$$

Equation 2.70 is now in a form similar to that of a linear Kalman filter.

The derivation of the EIF then proceeds by substituting Equations 2.68 and 2.69 into Equation 2.70.

$$\hat{\mathbf{x}}(k \mid k)$$
$$= [\mathbf{1} - \mathbf{W}(k)\nabla \mathbf{h}_x(k)]\hat{\mathbf{x}}(k \mid k-1) + \mathbf{W}(k)\mathbf{z}'(k)$$
$$= [\mathbf{P}(k \mid k)\mathbf{P}^{-1}(k \mid k-1)]\hat{\mathbf{x}}(k \mid k-1) + [\mathbf{P}(k \mid k)\nabla \mathbf{h}_x{}^T(k)\mathbf{R}^{-1}(k)]\mathbf{z}'(k).$$

Pre-multiplying both sides by $\mathbf{P}^{-1}(k \mid k)$ gives

$$\mathbf{P}^{-1}(k \mid k)\hat{\mathbf{x}}(k \mid k) = [\mathbf{P}^{-1}(k \mid k)\mathbf{P}(k \mid k)\mathbf{P}^{-1}(k \mid k-1)]\hat{\mathbf{x}}(k \mid k-1) +$$
$$[\mathbf{P}^{-1}(k \mid k)\mathbf{P}(k \mid k)\nabla \mathbf{h}_x{}^T(k)\mathbf{R}^{-1}(k)]\mathbf{z}'(k)$$
$$= \mathbf{P}^{-1}(k \mid k-1)\hat{\mathbf{x}}(k \mid k-1) + \nabla \mathbf{h}_x{}^T(k)\mathbf{R}^{-1}(k)\mathbf{z}'(k)$$
$$\Leftrightarrow \hat{\mathbf{y}}(k \mid k) = \hat{\mathbf{y}}(k \mid k-1) + \mathbf{i}(k). \tag{2.73}$$

This is the extended information estimation equation where the *information contribution* from nonlinear observation $\mathbf{z}(k)$ is given by

$$\mathbf{i}(k) = \nabla \mathbf{h}_x{}^T(k)\mathbf{R}^{-1}(k)\mathbf{z}'(k), \tag{2.74}$$

where $\mathbf{z}'(k)$ is the *'equivalent'* linearized observation given by Equation 2.71. The vector $\mathbf{z}'(k)$ gives an expression of the system observations if the nonlinear system is replaced by an equivalent linearized system. It depends on the innovation, the state prediction and the Jacobian evaluated at this prediction.

To compute the information matrix update, Equations 2.68 and 2.69 are substituted into the EKF variance update equation,

$$
\begin{aligned}
&\mathbf{P}(k \mid k) \\
&= [\mathbf{1} - \mathbf{W}(k)\nabla\mathbf{h}_x(k)]\mathbf{P}(k \mid k-1)[\mathbf{1} - \mathbf{W}(k)\nabla\mathbf{h}_x(k)]^T + \mathbf{W}(k)\mathbf{R}(k)\mathbf{W}^T(k) \\
&= [\mathbf{P}(k \mid k)\mathbf{P}^{-1}(k \mid k-1)]\mathbf{P}(k \mid k-1)[\mathbf{P}(k \mid k)\mathbf{P}^{-1}(k \mid k-1)]^T + \\
&\quad [\mathbf{P}(k \mid k)\nabla\mathbf{h}_x{}^T(k)\mathbf{R}^{-1}(k)]\mathbf{R}(k)[\mathbf{P}(k \mid k)\nabla\mathbf{h}_x{}^T(k)\mathbf{R}^{-1}(k)]^T. \quad (2.75)
\end{aligned}
$$

Pre- and post-multiplying both sides by $\mathbf{P}^{-1}(k \mid k)$ gives

$$
\Leftrightarrow \mathbf{P}^{-1}(k \mid k) = \mathbf{P}^{-1}(k \mid k-1) + \nabla\mathbf{h}_x{}^T(k)\mathbf{R}^{-1}(k)\nabla\mathbf{h}_x(k) \quad (2.76)
$$

$$
\Leftrightarrow \mathbf{Y}(k \mid k) = \mathbf{Y}(k \mid k-1) + \mathbf{I}(k). \quad (2.77)
$$

This is the linearized information matrix update equation where the associated matrix contribution is given by

$$
\mathbf{I}(k) = \nabla\mathbf{h}_x{}^T(k)\mathbf{R}^{-1}(k)\nabla\mathbf{h}_x(k). \quad (2.78)
$$

To obtain the corresponding prediction equations, consider the EKF state and variance prediction equations. Pre-multiplying the state prediction Equation 2.61 by $\mathbf{P}^{-1}(k \mid k-1)$ and inverting the variance prediction Equation 2.62 gives the information vector prediction as

$$
\hat{\mathbf{x}}(k \mid k-1) = \mathbf{f}\left(\hat{\mathbf{x}}(k-1 \mid k-1), \mathbf{u}(k-1), (k-1)\right)
$$

$$
\Leftrightarrow \mathbf{P}^{-1}(k \mid k-1)\hat{\mathbf{x}}(k \mid k-1) = \mathbf{P}^{-1}(k \mid k-1) \times
$$

$$
\mathbf{f}\left(\hat{\mathbf{x}}(k-1 \mid k-1), \mathbf{u}(k-1), (k-1)\right)
$$

$$
\Leftrightarrow \hat{\mathbf{y}}(k \mid k-1) = \mathbf{Y}(k \mid k-1)\ \mathbf{f}\left(\hat{\mathbf{x}}(k-1 \mid k-1), \mathbf{u}(k-1), (k-1)\right).
$$

The linearized information matrix prediction is obtained as follows:

$$
\mathbf{P}(k \mid k-1) = \nabla\mathbf{f}_x(k)\mathbf{P}(k-1 \mid k-1)\nabla\mathbf{f}_x{}^T(k) + \mathbf{Q}(k-1)
$$

$$
\Leftrightarrow \mathbf{Y}(k \mid k-1) = \left[\nabla\mathbf{f}_x(k)\mathbf{Y}^{-1}(k-1 \mid k-1)\nabla\mathbf{f}_x{}^T(k) + \mathbf{Q}(k)\right]^{-1}.
$$

This completes the derivation of the EIF; the entire algorithm can be summarized as follows:

### 2.5.3  Summary of the EIF Algorithm

**Prediction**

$$\hat{\mathbf{y}}(k \mid k-1) = \mathbf{Y}(k \mid k-1)\mathbf{f}\left(k, \hat{\mathbf{x}}(k-1 \mid k-1), \mathbf{u}(k-1), (k-1)\right) \quad (2.79)$$

$$\mathbf{Y}(k \mid k-1) = \left[\nabla \mathbf{f}_x(k)\mathbf{Y}^{-1}(k-1 \mid k-1)\nabla \mathbf{f}_x{}^T(k) + \mathbf{Q}(k)\right]^{-1}. \quad (2.80)$$

**Estimation**

$$\hat{\mathbf{y}}(k \mid k) = \hat{\mathbf{y}}(k \mid k-1) + \mathbf{i}(k) \tag{2.81}$$

$$\mathbf{Y}(k \mid k) = \mathbf{Y}(k \mid k-1) + \mathbf{I}(k). \tag{2.82}$$

The information state contribution and its associated information matrix are given, respectively, as

$$\mathbf{I}(k) = \nabla \mathbf{h}_x{}^T(k)\mathbf{R}^{-1}(k)\nabla \mathbf{h}_x(k) \tag{2.83}$$

$$\mathbf{i}(k) = \nabla \mathbf{h}_x{}^T(k)\mathbf{R}^{-1}(k)\left[\nu(k) + \nabla \mathbf{h}_x(k)\hat{\mathbf{x}}(k \mid k-1)\right], \tag{2.84}$$

where $\nu(k)$ is the innovation given by

$$\nu(k) = \mathbf{z}(k) - \mathbf{h}\left(\hat{\mathbf{x}}(k \mid k-1)\right). \tag{2.85}$$

### 2.5.4  Filter Characteristics

This filter has several attractive practical features, in particular:

- The filter solves, in information space, the linear estimation problem for systems with both nonlinear dynamics and observations. In addition to having all the attributes of the Information filter, it is a more practical and general filter.

- The information estimation Equations 2.81 and 2.82 are computationally simpler than the EKF estimation equations. This makes the partitioning of these equations for decentralized systems easy.

- Although the information prediction Equations 2.79 and 2.80 are of the same apparent complexity as the EKF ones, they are easier to distribute and fuse because of the *orthonormality* properties of information space parameters.

- Since the EIF is expressed in terms of information matrices and vectors, it is easily initialized compared to the EKF. Accurate initialization is important where linearized models are employed.

Some of the drawbacks inherent in the EKF still affect the EIF. These include the nontrivial nature of Jacobian matrix derivation (and computation) and linearization instability.

## 2.6   Examples of Estimation in Nonlinear Systems

In order to compare the extended Kalman filter and the extended Information filter, and illustrate the characteristics discussed above, three estimation problems in nonlinear systems are considered. These examples are chosen such that all possible combinations of nonlinearities in observations and nonlinearities in the system evolution are exhausted.

### 2.6.1   Nonlinear State Evolution and Linear Observations

Consider a two-dimensional radar tracking system which tracks a missile which is traveling vertically in an $xy$ plane with known vertical velocity $v$ and acceleration $a$, such that $\dot{x}(k) = 0$, $\dot{y}(k) = v$ and $\ddot{y}(k) = a$. The missile is fired vertically, in the positive $y$ axis direction, from some point on the $x$-axis. The radar is located at the origin of the $xy$ plane such that it measures the polar coordinates of the missile, that is, the radial position $r(k)$, and the angular displacement from the horizontal $\theta(k)$ where

$$r(k) = \sqrt{x^2(k) + y^2(k)} \quad and \quad \theta(k) = arctan\left[\frac{y(k)}{x(k)}\right].$$

Using the polar measurements, the objective is to estimate the entire missile state vector $\mathbf{x}(k)$ given by

$$\mathbf{x}(k) = \begin{bmatrix} x_1(k) \\ x_2(k) \\ x_3(k) \\ x_4(k) \end{bmatrix} = \begin{bmatrix} r(k) \\ \dot{r}(k) \\ \theta(k) \\ \dot{\theta}(k) \end{bmatrix}.$$

The equations of motion with respect to the polar coordinates, $r(k)$ and $\theta(k)$, are obtained by taking first and second derivatives. The result can be expressed in a nonlinear state vector form as follows:

$$\dot{\mathbf{x}}(k) = \begin{bmatrix} \dot{r}(k) \\ \ddot{r}(k) \\ \dot{\theta}(k) \\ \ddot{\theta}(k) \end{bmatrix} = \begin{bmatrix} v \sin\theta(k) \\ a\sin\theta(k) + \frac{v^2}{r(k)}\cos^2\theta(k) \\ \frac{v}{r(k)}\cos\theta(k) \\ \frac{ar(k) - v^2\sin\theta(k)}{r^2(k)}\cos\theta(k) - \frac{v^2}{r^2(k)}\sin\theta(k)\cos\theta(k) \end{bmatrix}.$$

Using the definition of the derivative of a generic state vector element $x_i(k)$ leads to

$$\dot{x}_i(k-1) = \frac{x_i(k) - x_i(k-1)}{\Delta T},$$

$$\Leftrightarrow x_i(k) = x_i(k-1) + \Delta T \dot{x}_i(k-1). \tag{2.86}$$

The discrete-time nonlinear system model is then obtained by employing Equation 2.86 for all elements of the state vector $\mathbf{x}(k)$, while using the state element derivatives from the vector $\dot{\mathbf{x}}(k)$ (assuming $\Delta T = 1$).

$$\mathbf{x}(k)$$
$$= \mathbf{f}\left(\mathbf{x}(k-1), (k-1)\right) + \mathbf{w}(k)$$

$$= \begin{bmatrix} x_1(k-1) + v \sin x_3(k-1) \\ x_2(k-1) + a \sin x_3(k-1) + \frac{v^2}{x_1(k-1)} \cos^2 x_3(k-1) \\ x_3(k-1) + \frac{v}{x_1(k-1)} \cos x_3(k-1) \\ x_4(k-1) + \frac{\left[ax_1(k-1) - v^2 \sin x_3(k-1)\right] \cos x_3(k-1) - v^2 \sin x_3(k-1) \cos x_3(k-1)}{x_1^2(k-1)} \end{bmatrix}$$

$$+\mathbf{w}(k).$$

The Jacobian of this nonlinear model is given by

$$\nabla \mathbf{f}(x) = \begin{bmatrix} a_{11} \ a_{12} \ a_{13} \ a_{14} \\ a_{21} \ a_{22} \ a_{23} \ a_{24} \\ a_{31} \ a_{32} \ a_{33} \ a_{34} \\ a_{41} \ a_{42} \ a_{43} \ a_{44} \end{bmatrix},$$

where

$$a_{11} = 1.0$$
$$a_{12} = 0.0$$
$$a_{13} = v \cos x_3(k-1)$$
$$a_{14} = 0.0$$
$$a_{21} = -\frac{v^2 \cos^2 x_3(k-1)}{x_1^2(k-1)}$$
$$a_{22} = 1.0$$
$$a_{23} = a \cos x_3(k-1) - \frac{v^2}{x_1(k-1)} \sin 2x_3(k-1)$$
$$a_{24} = 0.0$$
$$a_{31} = -\frac{v \cos x_3(k-1)}{x_1^2(k-1)}$$
$$a_{32} = 0.0$$
$$a_{33} = 1 - \frac{v \sin x_3(k-1)}{x_1(k-1)}$$

$a_{34} = 0.0$

$$a_{41} = \frac{\left[2v^2 x_1(k-1)\sin 2x_3(k-1)\right] - \left[ax_1{}^2(k-1)\cos x_3(k-1)\right]}{x_1^4(k-1)}$$

$a_{42} = 0.0$

$$a_{43} = -\left[\frac{ax_1(k-1)\sin x_3(k-1) + 2v^2 \cos 2x_3(k-1)}{x_1^2(k-1)}\right]$$

$a_{44} = 1.0.$

Since the radial position $r(k)$ and the angular displacement $\theta(k)$ are linearly measured by the radar, the observation matrix for the tracking system is given by

$$\mathbf{H}(k) = \begin{bmatrix} 1 & 0 & 0 & 0 \\ 0 & 0 & 1 & 0 \end{bmatrix}.$$

These system modeling matrices and vectors are then used in the algorithms of the EKF and EIF to carry out estimation of the missile state vector. The process and observation noises are generated by random number generators as explained in Section 2.3.4. The results are discussed in Section 2.6.4.

## 2.6.2   Linear State Evolution with Nonlinear Observations

A system may involve linear system dynamics and nonlinear measurement equations. An example of such a system is a radar station which makes measurements of the radial position $r(k)$ and the angular displacement $\theta(k)$ of an aircraft, from which it is desired to obtain the estimated values of the horizontal and vertical positions and velocities of the aircraft. The motion of the aircraft is such that the horizontal velocity $v_x$ and the vertical velocity $v_y$ are constant, which means the aircraft is executing linear motion. This is a four dimensional problem involving two positions and two velocities.

As in the previous radar and missile example (in Section 2.6.1) the polar coordinates and the Cartesian coordinates are related by the equations

$$r(k) = \sqrt{x^2(k) + y^2(k)} \quad and \quad \theta(k) = arctan\left[\frac{y(k)}{x(k)}\right].$$

The state vector of interest consists of the two positions and the two velocities such that

$$\mathbf{x}(k) = \begin{bmatrix} x_1(k) \\ x_2(k) \\ x_3(k) \\ x_4(k) \end{bmatrix} = \begin{bmatrix} x(k) \\ y(k) \\ \dot{x}(k) \\ \dot{y}(k) \end{bmatrix} = \begin{bmatrix} v_x k \\ v_y k \\ v_x \\ v_y \end{bmatrix}.$$

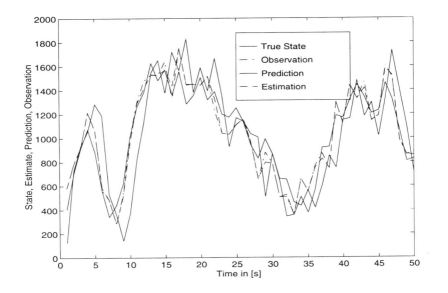

**FIGURE 2.6**
**EKF & EIF: Nonlinear State Evolution with Linear Observations**

The system models are established as in previous examples. In particular,

$$\mathbf{F}(k) = \begin{bmatrix} 1 & 0 & \Delta T & 0 \\ 0 & 1 & 0 & \Delta T \\ 0 & 0 & 1 & 0 \\ 0 & 0 & 0 & 1 \end{bmatrix}$$

$$\mathbf{z}(k) = \begin{bmatrix} r(k) \\ \theta(k) \end{bmatrix} + \mathbf{v}(k)$$

$$= \mathbf{h}\left(\mathbf{x}(k), k\right) + \mathbf{v}(k)$$

$$= \begin{bmatrix} \sqrt{x_1{}^2(k) + x_2{}^2(k)} \\ \arctan\left[\frac{x_2(k)}{x_1(k)}\right] \end{bmatrix} + \mathbf{v}(k)$$

$$\nabla \mathbf{h}_x(k) = \begin{bmatrix} \frac{x_1(k)}{\sqrt{x_1^2(k)+x_2^2(k)}} & \frac{x_2(k)}{\sqrt{x_1^2(k)+x_2^2(k)}} & 0 & 0 \\ \frac{-x_2(k)}{x_1^2(k)+x_2^2(k)} & \frac{x_1(k)}{x_1^2(k)+x_2^2(k)} & 0 & 0 \end{bmatrix}.$$

These system models are then used in the algorithms of the EKF and EIF

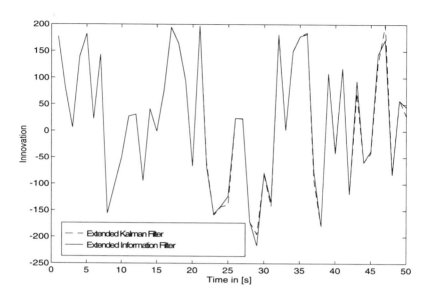

**FIGURE 2.7**
**Innovations for EKF and EIF: Nonlinear State Evolution with Linear Observations**

to estimate the state vector of the aircraft's horizontal and vertical positions and velocities. The results are also presented and discussed in Section 2.6.4.

### 2.6.3    Nonlinear State Evolution with Nonlinear Observations

A highly nonlinear system has nonlinearities in both system state evolution and observations. An example of such a system is a wheeled mobile robot (WMR) vehicle moving in a plane. The state vector of the vehicle at any time instant $k$ is determined by its location and orientation such that

$$\mathbf{x}(k) = \begin{bmatrix} x(k) \\ y(k) \\ \phi(k) \end{bmatrix},$$

where $x(k)$ and $y(k)$ denote the WMR positions along the $x$ and $y$ axes of the plane, respectively, and $\phi(k)$ is the WMR orientation. Control is extended over the WMR vehicle motion through a demanded velocity $v(k)$ and direction of travel $\psi(k)$,

$$\mathbf{u}(k) = \begin{bmatrix} v(k) \\ \psi(k) \end{bmatrix}.$$

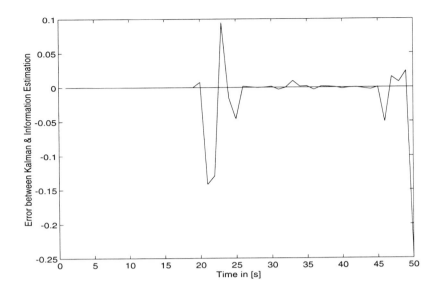

**FIGURE 2.8**
**Difference between EKF and EIF State Estimates: Nonlinear State Evolution with Linear Observations**

The motion of the vehicle can now be described in terms of the simple nonlinear state transition equation,

$$
\begin{bmatrix} x(k) \\ y(k) \\ \phi(k) \end{bmatrix} = \begin{bmatrix} x(k-1) + \Delta T v(k) \cos\left[\phi(k-1) + \psi(k)\right] \\ y(k-1) + \Delta T v(k) \sin\left[\phi(k-1) + \psi(k)\right] \\ \phi(k-1) + \Delta T \frac{v(k)}{B} \sin\psi(k) \end{bmatrix} + \begin{bmatrix} w_x(k) \\ w_y(k) \\ w_\phi(k) \end{bmatrix},
$$

where $B$ is the wheel base line, $\Delta T$ is the time in travel between time steps, and $\mathbf{w}(k) = [w_x(k), w_y(k), w_\phi(k)]^T$ is the random vector describing the noise in the process due to both modeling errors and uncertainty in control. It is assumed that the vehicle is equipped with a sensor that can measure the range and bearing to a moving beacon with motion described by two parameters, $B_i = [x_i, y_i]^T$, such that $x_i$ varies linearly with time, i.e., $x_i = 0.5k$. Assuming that the beacon is moving in circular motion of radius 10 units about the vehicle, then $y_i$ is given by the expression, $y_i = \sqrt{100 - x_i^2}$. The observation equations for the beacon are given by

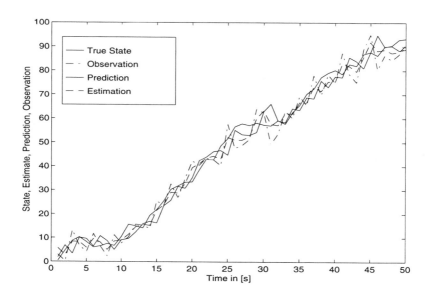

**FIGURE 2.9**
**EKF & EIF: Linear State Evolution with Nonlinear Observations**

the nonlinear measurement model,

$$
\begin{bmatrix} z_r{}^i(k) \\ z_\theta{}^i(k) \end{bmatrix} = \begin{bmatrix} \sqrt{[x_i - x(k)]^2 + [y_i - y(k)]^2} \\ \arctan\left[\frac{y_i - y(k)}{x_i - x(k)}\right] - \phi(k) \end{bmatrix} + \begin{bmatrix} v_r(k) \\ v_\theta(k) \end{bmatrix},
$$

where the random vector $\mathbf{v}(k) = [v_r(k), v_\theta(k)]^T$ describes the noise in the observation process. The system models are defined and established as before. In particular,

$$
\nabla \mathbf{f_x}(k) = \begin{bmatrix} 1 & 0 & -\Delta T v(k) \sin\left[\hat{\phi}(k-1 \mid k-1) + \psi(k)\right] \\ 0 & 1 & \Delta T v(k) \cos\left[\hat{\phi}(k-1 \mid k-1) + \psi(k)\right] \\ 0 & 0 & 1 \end{bmatrix}
$$

$$
\nabla \mathbf{h_x}(k) = \begin{bmatrix} \frac{\hat{x}(k|k-1)-x_i}{d} & \frac{\hat{y}(k|k-1)-y_i}{d} & 0 \\ -\frac{\hat{y}(k|k-1)-y_i}{d^2} & \frac{\hat{x}(k|k-1)-x_i}{d^2} & -1 \end{bmatrix},
$$

where $d = \sqrt{[x_i - \hat{x}(k|k-1)]^2 + [y_i - \hat{y}(k|k-1)]^2}$.

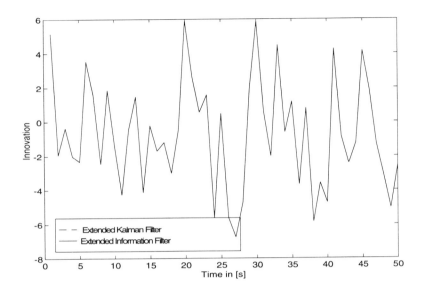

**FIGURE 2.10**
**Innovations for EKF and EIF: Linear State Evolution with Non-linear Observations**

These system models are then used in the algorithms of the EKF and EIF to estimate the WMR vehicle's state vector $\mathbf{x}(k)$, that is, estimate the location and orientation of the vehicle. The results are also presented and discussed in Section 2.6.4.

## 2.6.4  Comparison of the EKF and EIF

The EKF and EIF are compared in the same way as the linear filters. However, the three nonlinear estimation examples outlined in Sections 2.6.1, 2.6.2 and 2.6.3 are implemented for each filter. These examples were chosen to allow exhaustive investigation of nonlinearities in both system evolution and observations. In order to study and compare the performance of the filters, for each example, estimation of the same state is considered.

The general filter performance for these examples is shown in Figures 2.6 and 2.9. As was the case with the linear filters, the state observations, predictions and estimates are identical for the EKF and EIF. The curves depicting the same variables are indistinguishable because they lie on top of each other. The same equivalence is observed for estimation errors and the innovations. There are still slight differences between the two filters, attributable to numerical errors. As the amount of nonlinearities increases,

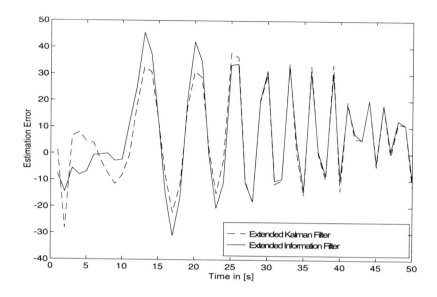

**FIGURE 2.11**
**Estimation Errors for EKF and EIF: Nonlinear State Evolution
with Nonlinear Observations**

the errors tend to increase. However, even in the worst case, the errors are
still bounded and non-consequential.

The nature and trend of the difference between the EKF and EIF state
estimates is shown in Figures 2.8 and 2.13. The errors are worse than
those for linear systems because of the need to compute the Jacobians,
$\nabla \mathbf{f}_x(k)$ and $\nabla \mathbf{h}_x(k)$. The Jacobians are not constants; they are functions
of both time-step and state. As a result the covariances and system models
must be computed on-line. This increases the amount of computations performed and hence the numerical errors between the two filters are greater.
The greater the complexity of the nonlinearities, the greater the number
and complexity of the Jacobians, which leads to more computational costs.
This tends to produce more numerical and rounding off errors. In spite of
these errors, the equivalence of the EKF and EIF is amply demonstrated
in the three examples. This confirms the algebraic equivalence which is
mathematically proven and established in the derivation of the EIF from
the EKF and the Information filter.

In terms of filter performance, both the EKF and EIF filters show unbiasedness, consistency, efficiency and good matching. In all three examples
the state estimate is always well placed between the observation and state
prediction. This means there is balanced confidence in observations and

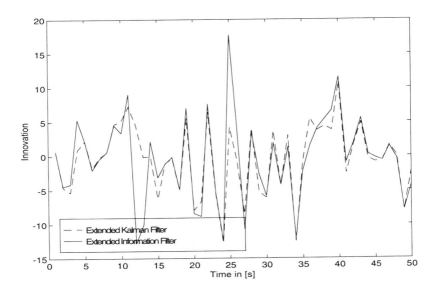

**FIGURE 2.12**
**Innovations for EKF and EIF: Nonlinear State Evolution with Nonlinear Observations**

predictions. By inspection and computing the sequence mean, the innovations (Figures 2.7, 2.10 and 2.12) are shown to be zero mean with variance $\mathbf{S}(k)$. There is no visible correlation of the innovations sequences and they satisfy the 95% confidence rule. However, in general, the performance of the EKF and EIF is not as good as that of the linear filters. This is because of the nontrivial nature of the Jacobian matrix computation and the general instability inherent in linearized filters.

## 2.7 Summary

This chapter has developed estimation techniques which form the basis of the decentralized estimation and control presented in this book. The notation and system description have been introduced and explained. Estimation theory and its use were discussed, in particular, the Kalman filter algorithm was outlined. The information filter was then derived as an algebraic equivalent to the traditional Kalman filter. Its attributes were outlined and discussed. The extended Kalman filter was then presented as

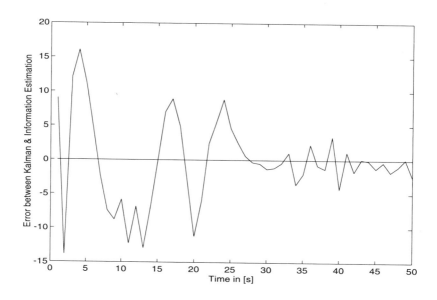

**FIGURE 2.13**
**Difference between EKF and EIF State Estimates: Nonlinear**
**State Evolution with Nonlinear Observations**

a state space solution to the estimation problem for a system characterized by both nonlinear system evolution and nonlinear measurements. The
original and novel contribution of this chapter is the extended Information
filter, EIF. This algorithm provides an estimation technique in extended information space for nonlinear systems. It was derived from first principles,
explained and appraised. It has all the attributes of the linear Information filter and less of the problems associated with the EKF. The simulated
examples of estimation in linear and nonlinear systems, validated the Information filter and EIF algorithms with respect to those of the Kalman filter
and EKF. For the EIF and EKF, examples involving nonlinearities in both
system evolution and observations were considered. The key benefit of information estimation theory is that it makes fully decentralized estimation
for multisensor systems (developed in Chapter 3) attainable.

# Chapter 3

## Decentralized Estimation for Multisensor Systems

### 3.1 Introduction

This chapter addresses the multisensor estimation problem for both linear and nonlinear systems in a fully connected decentralized sensing architecture. The starting point is a brief review of sensor characteristics, applications, classification and selection. An overview of multisensor systems and their attributes is presented in order to provide an understanding of the background to, and motivation for multisensor systems. The sensor data fusion problem is then identified and literature addressing it is discussed while the issue of fusion architectures is considered. Three main categories of architectures: centralized, hierarchical and decentralized are presented and appraised. This serves the purpose of explaining the advantages of decentralization. A working definition for a decentralized system is then proposed and the benefits of such a system outlined. A brief survey of literature covering decentralized systems is then presented.

The remainder of this chapter develops fully connected decentralized estimation algorithms in both state and information spaces. The starting point is partitioning the linear observation models to produce a decentralized observer while retaining a central estimator. By further decentralizing the information form of the Kalman filter prediction equations, and by communicating partial information estimates, a decentralized Information filter is obtained. A state space representation of the same algorithm is outlined. The intent is to show that decentralized estimation is feasible and to demonstrate the advantages of information space over state space. The decentralization procedure is then repeated for the EKF and EIF to produce decentralized filters for nonlinear systems. The problems associated with a fully connected topology are outlined, thus setting the stage for Chapter 4, which seeks to remove this requirement.

## 3.2   Multisensor Systems

The algorithms presented in Chapter 2 are estimators for single sensor systems. A sensor is a device which receives a signal or stimulus and generates measurements that are functions of that stimulus. Most sensors consist of a transducer and an electronic circuit, where the transducer converts a physical or chemical quantity to an electrical signal, as exemplified by a strain gauge and a solar cell, respectively. The use of sensors is pervasive in many fields, for example, in robotics sensors are used to measure the location of the robot (localization problem), finding location of objects, detecting and avoiding obstacles, monitoring interaction with the environment, measuring and correcting modeling errors, monitoring changes in the environment and obtaining parameters required to control the robot.

### 3.2.1   Sensor Classification and Selection

Sensor classification schemes range from very simple to the complex. One approach is to consider all of the sensor's properties such as the stimulus it measures, its specifications, the physical phenomenon it is sensitive to, the conversion mechanism it employs, the material it is fabricated from and its field of application [44]. In another scheme sensors are classified according to five categories: *state, function, performance, output and energy type* [26].

- **State**: In this category sensors are classified as either internal or external state sensors. Internal state sensors are devices used to measure system parameters such as position, velocity and acceleration. Examples of such sensors include potentiometers, tachometers, accelerometers and optical encoders. External state sensors are used to monitor the system's geometric and/or dynamic relation to its tasks and environment [65]. Examples of such sensors include proximity devices, strain gauges, sonar (ultrasonic range sensor), pressure sensors and electromagnetic sensors.

- **Function**: Sensors are also classified in terms of the parameters which they measure. In the mechanical domain such measurands include displacement (linear and angular), velocity (linear, angular and flow rate), acceleration (vibration and shock), dimensional (position, size, volume and strain), mass (weight, load and density) and force (absolute, relative, static, torque and pressure). Other types of sensor function are hardness and viscosity.

- **Performance**: Measures of performance can also be used to classify sensors. These measures include accuracy, repeatability, linearity, sensitivity, resolution, reliability and range. The selection of an appropriate sensor device requires that the available devices be examined against each of the relevant performance parameters [26].

- **Output**: The type of sensor output is also useful as a criterion of classification, and output signals fall into four general categories: analogue (a continuous output signal), digital (digital representation of measurand), frequency (use of output signal's frequency) and coded (modulation of output signal's frequency, amplitude or pulse).

- **Energy Type**: This classification is based on the type of energy transfer in the transducer. Radiant energy is involved in electromagnetic radiation, frequency, phase and intensity, while mechanical energy involves such mechanical measurands as distance, velocity and size. Thermal energy covers the measurement of temperature effects in materials including thermal capacity, latent heat and phase change properties, whereas electrical energy deals with electrical parameters such as current, voltage, resistance and capacitance. The other two energy types are magnetic and chemical.

All sensors may also be put into two general categories depending on whether they are *passive* or *active* [44], [65].

- **Passive Sensors**: These sensors directly generate an electric signal in response to an external stimulus, i.e., the input stimulus energy is converted by the sensor into output energy without the need for an additional power source or *injection* of energy into the sensor. Examples of such sensors are a thermocouple, a piezoelectric sensor, a pyroelectric detector and a passive infrared sensor (PIF) which is shown in Figure 3.1.

- **Active Sensors**: In contrast to passive sensors, active sensors require external power for their operation, which is called an *excitation signal*. This means the sensor *injects* energy into the system which is being monitored, and the signal being measured is thus modified by the sensor to produce an output signal. Active sensors are also called *parametric* sensors because their own properties change in response to an external effect and these properties are subsequently converted into electric signals. For example, a thermistor is a temperature measuring device which consists of a resistor whose resistance changes with temperature. When an electric current (excitation signal) is passed through the thermistor, its change in resistance is obtained by detecting the variation in the voltage across the thermistor. This change in voltage is related to the temperature being measured.

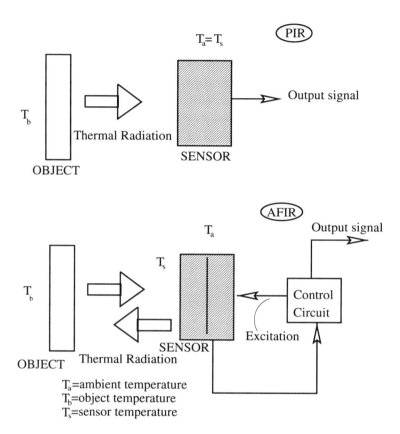

**FIGURE 3.1**
**Passive and Active Sensors: PIR and AFIR**

Other examples of active sensors are an active far infrared sensor
(AFIR) (compared with a PIR sensor in Figure 3.1), an active optical
seam tracker and an active computer vision system [44].

Although these different ways of classifying sensors are overlapping and
not exhaustive they are useful in sensor selection, which is the process of
choosing the best sensor for a specific application. Quite often, the same
stimulus may be measured by using different physical phenomena, and subsequently by different sensors. It is therefore a matter of an engineering
choice to select the best sensor for a particular task. Selection criteria depend on many factors such as availability, cost, power consumption, nature
of stimulus and environmental conditions [44].

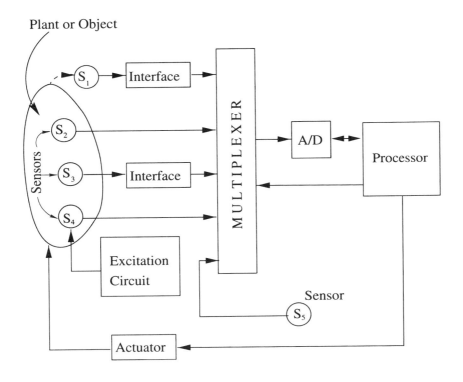

**FIGURE 3.2**
**Positions of Sensors in a Data Acquisition System**

## 3.2.2 Positions of Sensors in a Data Acquisition System

A sensor does not function by itself, it is always a part of a larger system, the data acquisition system, which may incorporate many other sensors, signal conditioners, signal processors, memory devices, data recorders and actuators. Figure 3.2 illustrates the positions of various sensors in a typical data acquisition system [44]. Some of the sensors (2, 3 and 4) are positioned directly on or inside the plant (object) being monitored and are called *contact* sensors. Sensors that measure the object without a physical contact, as is the case with sensor 1, are called *noncontact* sensors. Examples of such sensors are a radiation detector and a TV camera.

Sensor 5 is a special sensor that monitors the internal conditions of the acquisition system itself. Some sensors (1 and 3) cannot be directly connected to standard electronic circuits due to inappropriate output signal formats and hence require the use of interface devices (signal conditioners).

Sensors 1, 2, 3 and 5 are passive while sensor 4 is active, which means it requires an operating signal which is provided by an excitation circuit. The number of sensors in the data acquisition system depends on the complexity of the system. For example, one sensor is required in a thermostat while thousands are required in a space shuttle.

Electrical signals from the sensors are fed into a multiplexer (MUX), which is a switch or a gate. Its function is to connect sensors to an analog-to-digital converter (A/D) or directly to a main processor (for a sensor that produces digital signals). The processor controls the timing of the multiplexer and the A/D, and also sends control signals to actuators which act on the plant (object) being monitored. Such actuators could be electric motors, solenoids, relays and pneumatic valves. The data acquisition system also contains some peripheral devices such as data recorders, displays, alarms, amplifiers and sample-and-hold circuits [44].

### 3.2.3   The Advantages of Multisensor Systems

In a single sensor system one sensor is selected to monitor the system or its surrounding environment. However, many advanced and complex applications require large numbers of sensors, rendering single sensor systems inadequate. A multisensor system employs several sensors to obtain information in a real world environment full of uncertainty and change. This means various types of sensors and different sensor technologies are employed, where some of these sensors have overlapping measurement domains. Multiple sensors provide more information and hence a better and more precise understanding of a system. Moreover, a single sensor is not capable of obtaining all the required information reliably at all times in varying environments. Furthermore, as the size and complexity of a system increases, so does the number and diversity of sensors required to capture its description. These are the primary motivating issues behind multisensor systems. There is a considerable amount of literature on the limitations of single sensor systems and the merits of multisensor systems [48], [69], [71], [80]. Multisensor systems have found applications in process control, robotics, navigation, aerospace and defense systems.

The advantages of multisensor systems include the following:

- Failure of a single sensor does not mean *complete* failure of the entire system because the other sensors can continue to be used. Overlap between sensor domains gives the system some degree of redundancy. Consequently, when sensor failure occurs, the system undergoes graceful degradation and not catastrophic failure.

- Different types of sensors can be used to give a more complete picture of the environment. Thus, different sensor technologies are utilized in the same application to provide improved system performance.

- Erroneous readings from a single sensor do not necessarily have a drastic effect on the system since information about the same environment can be obtained from other sensors. This property is particularly reinforced when there is extensive overlapping of sensor domains.

- Geographical diversity is provided by information from sensors placed at different positions in the sensed environment.

- Sensor selection is more flexible and effective as several sensors can be selected to monitor one specific task in a system. Thus, cheaper, redundant and complimentary sensors can be chosen as opposed to a single expensive sensor, while retaining the same reliability and increasing survivability.

### 3.2.4 Data Fusion Methods

In order for the advantages of multisensor systems to be realized, it is essential that the information provided by the sensors is interpreted and *combined* in such a way that a reliable, complete and coherent description of the system is obtained. This is the *data fusion* problem. Multisensor fusion is the process by which information from many sensors is combined to yield an improved description of the observed system. Fusion methods can either be *quantitative, qualitative* or a hybrid of both. Quantitative methods are based on numerical techniques while qualitative ones are based on symbolic representation of information. Examples of quantitative methods include statistical decision theory, identification techniques and probabilistic theory. Qualitative methods include expert systems, heuristics, behavioral and structural modeling. Several fusion approaches have been developed and applications effected in such areas as computer vision, robotics, diffuse systems and process control [3], [42], [68], [118].

In this book fusion based on probabilistic and statistical methods is employed. The approach consists of distributing and decentralizing the estimation algorithms derived in Chapter 2, allowing communication between sensor nodes and local assimilation of information. The principal estimation technique is information filtering based on the Kalman filter. This research is a further development to the work carried out by Berg [20], Grime [48] and Rao [106].

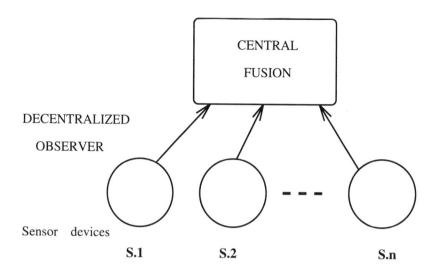

**FIGURE 3.3**
**Centralized Architecture**

### 3.2.5   Fusion Architectures

The taxonomy of fusion architectures corresponding to different fusion algorithms can be reduced to three general categories: *centralized, hierarchical* and *decentralized*. Detailed descriptions of these architectures and their advantages and limitations may be found in the literature [41], [80]. In this section only introductory notes and illustrative diagrams are presented.

**Centralized Architectures**

A fully centralized multisensor system comprises a central processor with direct connections to all sensor devices. Each of these devices obtains data about the environment which is forwarded to the central processor. The central processor is responsible for collecting readings from the sensor devices and processing the information obtained. Figure 3.3 illustrates a centralized fusion system. Conceptually, the algorithms used are similar to those for single sensor systems and hence relatively simple. Resource allocation is easy because the central processor has an overall view of the system. The central processor makes decisions based on the maximum possible information from the system. Since the central processor is fully aware of the information from each sensor and its activities, there should be no possibility of task or fusion duplication. Although centralized multisensor systems are an improvement on single sensor systems, they have a number of disadvantages. These include severe computational loads imposed on the

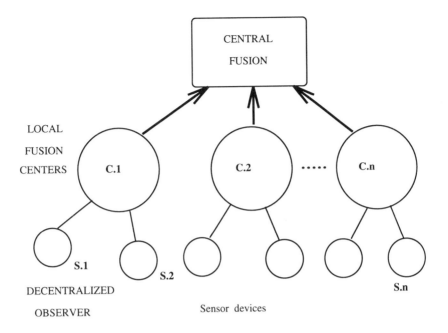

**FIGURE 3.4**
**Hierarchical Architecture**

central processor, the possibility of catastrophic failure (due to failure of the central node), high communication overheads and inflexibility to changes of application or sensor technology.

**Hierarchical Architectures**

A typical hierarchical architecture is shown in Figure 3.4. The principle of a hierarchy is to reduce the communication and computational problems of centralized systems by distributing data fusion tasks amongst a hierarchy of processors. In a hierarchy there is still a central processor acting as a fusion center. Processors constituting local fusion centers, locally process information and send it to the central processor. Extensive use of such systems has been made in robotics and surveillance applications. In fact, most advanced systems today are generally variants of hierarchical structures [59]. Although these systems have the advantage of distributing the computational load, they still retain some of the disadvantages associated with the centralized model. In addition to these problems, they have more implementational drawbacks which include algorithm requirements for sensor level tracking and data fusion, and vulnerability to communication bottlenecks.

### Decentralized Architectures

Most of the drawbacks of centralized and hierarchical architectures are resolved by using a fully decentralized architecture. The advantages of fully decentralized systems provide the basis and motivation for the estimation and control algorithms developed in this book. Consequently, such systems are formally defined and discussed in the next section.

## 3.3    Decentralized Systems

A decentralized system is a data processing system in which all information is processed locally and where there is no central processing site. It consists of a network of nodes, each with its own processing facility, which together do not require any central fusion or communication facility. In such a system, fusion occurs locally at each node on the basis of local observations and information communicated from neighboring nodes. At no point is there a common place where global decisions are made. Only node-to-node communication and local system knowledge is permitted.

For a decentralized data fusion (estimation) system, the processing node is a sensor (fusion) node which takes local observations and shares information with other fusion nodes. It then assimilates the communicated information and computes a local estimate. In decentralized control systems, in addition to fusion, the node uses the fused information to generate a local control signal. This is developed and discussed in Chapter 5. A general decentralized estimation system with arbitrary topology is shown in Figure 3.5. It consists of a set of communicating sensor nodes, none of which is central to the system. A typical sensor node is shown in Figure 3.6. The system monitor is used for extracting information from the system. It is arbitrarily located and hence no assumptions are made about its position in the network [107], [108], [115]. A decentralized architecture overcomes many of the problems associated with centralized and hierarchical systems.

### 3.3.1    The Case for Decentralization

Fully decentralized systems have several advantages, in particular the following characteristics motivate the work presented in this book:

- **Modularity:** As all fusion processes must take place locally, and no global knowledge of the network is required *a priori*, nodes can be

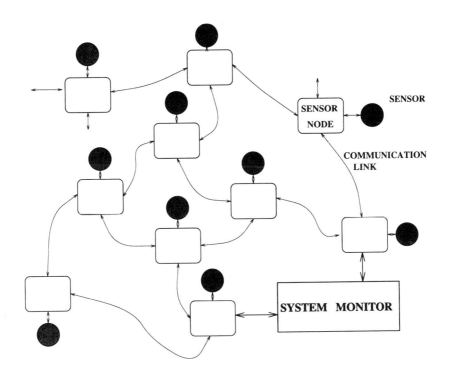

**FIGURE 3.5**
**Decentralized Architecture**

constructed and programmed in a modular fashion. Modularity is useful in system design, application and maintenance.

- **Survivability and Robustness:** Since system performance is not dependent on any one processor, it fails gracefully under node communication link failure. It is thus highly survivable and robust.

- **Flexibility:** Because no node is central and no global knowledge of the network topology is required, the system is flexible to on-line addition or loss of nodes and to dynamic changes in the network structure.

- **Extensibility:** Eliminating the central fusion center and any common communication facility makes the system scalable by reducing the limits imposed by centralized computational bottlenecks and lack of communication bandwidth.

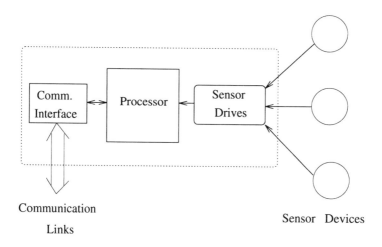

**FIGURE 3.6**
**Sensor Node**

These benefits apply to both decentralized estimation and control systems. However, there are associated drawbacks. In fully decentralized systems, communication issues are more complex. Communication overheads are higher than in centralized systems but less than in hierarchical ones. The number of processors required per system is higher. System monitoring is harder than in both centralized and hierarchical systems.

## 3.3.2    Survey of Decentralized Systems

In this book, the term decentralized has the specific definition given at the beginning of Section 3.3. This is not the same definition of decentralized used in parts of the literature. The term is often applied to any system comprising multiple estimators or controllers.

Work covering systems described as decentralized has been published in the area known as Large Scale Systems (LSS) or Complex Systems. Ho [57] defines a large scale system as a group of subsystems that are interconnected in such a way that decentralized operation is mandatory. The main characteristics of such systems include multiple decision makers, correlated but different information, coordination between nodes, imprecise models, conflict and multiple objectives. Work in LSS is then concerned with generating decentralized and distributed estimators and controllers for these systems. The research includes efforts to apply LSS ideas in such areas as economics, organizational theory and systems theory. Decentralized and hierarchical control methods for LSS are surveyed further by Sandell and Athans [113]. Of the work surveyed, the system which is closest to the definition given at the beginning of Section 3.3 consists of interacting controllers with a central processing site. Much of the work in the literature concentrates on interconnected dynamical subsystems and deriving transformations between their state spaces. However, these efforts do not solve the problem of fusing estimates from different state subspaces.

Siljak in his book [115] discusses a number of issues in decentralized control of complex systems: optimality, robustness, structural controllability, observability and 'essential uncertainties' (which cannot be described in either deterministic or stochastic terms). The book considers a decentralized system as an interacting network of subsystems, each with its own plant and controller. He concludes that extending standard optimization concepts to such systems is not possible and argues that robustness is part of the problem which should be considered from the onset. However, the systems in Siljak's research do not satisfy the definition of decentralization used in this book. Moreover, in his work the system optimality properties obtained decentrally are different from those obtained centrally.

In his papers [55], [56], Ho discusses team decision theory and the notion of information structure as approaches to address the LSS problem. The paper by Hashemipour [50] describes various multisensor networks whose signal processing tasks are amenable to multiprocessor implementation. The main estimation algorithm is the Kalman filter and the proposed network is a hierarchical or two-tier Kalman filter structure. The main principle is that the global estimate is centrally reconstructed from estimates computed by local nodes. The recent paper by Chong and Mori [38] discusses the non-fully connected problem in estimation. The formulation of the nodal estimators is based on an information form of the Kalman filter. In addition the paper describes a model for communication in a sensor network in terms of information graphs. There is expensive propagation of information in the networks and the issue of system nonlinearities is not addressed.

The work of Speyer [117] and, more recently, that of Rao and Durrant-Whyte [106] satisfy the definition of decentralization employed in this book. This work shows how estimates can be fused both optimally and decentrally. Speyer's paper derives a solution to the decentralized control problem and discusses the computation and communication requirements in a decentralized architecture. A control problem involving an arbitrary number of control nodes is formulated such that local control is generated using the best state estimate given information from all the nodal sensors. He shows that the conditions for stabilizing decentralized controllers are equivalent to those required for a global controller; however, he does not explicitly solve the decentralized estimation problem and does not exploit the use of information variables. Rao and Durrant-Whyte explicitly derived and implemented the decentralized Kalman filter. Further work based on their efforts has addressed several aspects of decentralized estimation, sensor management [70], communication [46], model distribution [20], system organization [54] and modular vehicle design [33].

However, this work does not address the extension of decentralized estimation to sensor based decentralized control, nor does it adequately address the problem of nonlinearities in system and observation models. With the exception of the work of Berg [20], there is no model distribution and the estimation topologies are fully connected. Fully connected topologies have very limited practical use for large systems. In the work of Berg [20], although model distribution is used the method is not sufficiently generalized, and the issue of how nodal transformation matrices are determined is not addressed. Non-fully connected estimation topologies presented in [46] and [54] require expensive propagation of information between unconnected nodes using a channel filter.

The decentralized estimation and control methods presented in this book are motivated by the contributions and associated limitations of the work surveyed above. This is evident from the constraints outlined in the problem statement given in Chapter 1.

## 3.4   Decentralized Estimators

The major advantage of information space is its structural and computational simplicity which makes it applicable to decentralized estimation. The estimation and prediction equations of the Information filter can easily be decentralized into a network of nodes communicating simple information components. With this approach the entire estimation process is carried out in terms of information about states and not the states themselves. In this section, decentralized estimation algorithms in information space are derived for both linear and nonlinear systems. The state space equivalents of these algorithms are also outlined in state space.

### 3.4.1   Decentralizing the Observer

The starting point in developing decentralized estimation algorithms is to partition the observation model and the corresponding observation equations to produce a decentralized observer. A system comprising $N$ sensors with a composite observation model is considered. The observation vector $\mathbf{z}(k)$ is unstacked into $N$ subvectors of dimension $N_i$ corresponding to the observations made by each individual sensor,

$$\mathbf{z}(k) = \left[\mathbf{z}_1^T(k), \cdots, \mathbf{z}_N^T(k)\right]^T. \tag{3.1}$$

The observation matrix and noise vector are then partitioned into submatrices and subvectors corresponding to these observations,

$$\mathbf{H}(k) = \left[\mathbf{H}_1^T(k), \cdots, \mathbf{H}_N^T(k)\right]^T, \tag{3.2}$$

$$\mathbf{v}(k) = \left[\mathbf{v}_1^T(k), \cdots, \mathbf{v}_N^T(k)\right]^T. \tag{3.3}$$

It is assumed that the noise vector partitions are uncorrelated,

$$E[\mathbf{v}(k)\mathbf{v}^T(k)] = \mathbf{R}(k) = blockdiag\{\mathbf{R}_1(k), \cdots, \mathbf{R}_N(k)\}. \tag{3.4}$$

As a result the sensor model now consists of $N$ equations in the form

$$\mathbf{z}_j(k) = \mathbf{H}_j(k)\mathbf{x}(k) + \mathbf{v}_j(k), \tag{3.5}$$

with $\mathbf{v}_j(k)$ modeled as an uncorrelated white sequence with local measurement noise covariance,

$$E[\mathbf{v}_j(k)\mathbf{v}_j^T(l)] = \delta_{kl}\mathbf{R}_j(k). \tag{3.6}$$

With this decentralized observer it might be tempting to try to integrate multiple observations into a single estimate using the usual form of the Kalman filter. This can be attempted by applying the same gain equation to each individual observation as

$$\hat{\mathbf{x}}(k \mid k) = \hat{\mathbf{x}}(k \mid k-1) + \sum_{i=1}^{N} \mathbf{W}_i(k)\left[\mathbf{z}_i(k) - \mathbf{H}_i(k)(\hat{\mathbf{x}}(k \mid k-1))\right] \tag{3.7}$$

$$\mathbf{W}_i(k) = \mathbf{P}(k \mid k-1)\mathbf{H}_i^T(k)\mathbf{S}_i^{-1}(k), \tag{3.8}$$

and

$$\mathbf{S}_i(k) = \mathbf{H}_i(k)\mathbf{P}(k \mid k-1)\mathbf{H}_i^T(k) + \mathbf{R}_i(k). \tag{3.9}$$

However, this cannot be done as the innovations between different sensors at any time-step are correlated. As a result of this correlation, the innovation covariance $\mathbf{S}(k)$ is not block diagonal, whereas this property is a requirement for the conventional form of the Kalman filter to be decoupled. This is the reason why the conventional form of the Kalman filter cannot be used to obtain a multisensor estimate from a linear combination of separate observations and models.

## 3.4.2 The Decentralized Information Filter (DIF)

The decentralized observer is used with the Information filter to provide a decentralized estimation algorithm, the decentralized Information filter (**DIF**). At this stage it is assumed that each local fusion node has a state space model identical to an equivalent centralized model. The issue of model

distribution, in which each node maintains only a part of the global model
is addressed in Chapter 4. Information from observations $\mathbf{z}_j(k)$ made at a
general node $j$ is defined in terms of a local information state contribution
$\mathbf{i}_j(k)$ and its associated information matrix $\mathbf{I}_j(k)$. These local information
parameters are given by the following equations:

$$\mathbf{i}_j(k) \triangleq \mathbf{H}_j^T(k)\mathbf{R}_j^{-1}(k)\mathbf{z}_j(k) \tag{3.10}$$

and

$$\mathbf{I}_j(k) \triangleq \mathbf{H}_j^T(k)\mathbf{R}_j^{-1}(k)\mathbf{H}_j(k). \tag{3.11}$$

Consider a general network of fully connected sensor nodes $N$ and let
$i$ and $j$ be any two nodes. The algorithm is derived for a general node $i$
with information being communicated to it from the other $(N-1)$ nodes
represented by $j$.

## Prediction

Each node starts by computing local predictions based on previous, locally determined information estimates and system models. The local system models are the same as the global ones because the system is fully
connected. Hence, the prediction equations have the form

$$\hat{\mathbf{y}}_i(k \mid k-1) = \mathbf{L}_i(k \mid k-1)\hat{\mathbf{y}}_i(k-1 \mid k-1) \tag{3.12}$$

$$\mathbf{Y}_i(k \mid k-1) = \left[\mathbf{F}(k)\mathbf{Y}_i^{-1}(k-1 \mid k-1)\mathbf{F}^T(k) + \mathbf{Q}(k)\right]^{-1}. \tag{3.13}$$

The local propagation coefficient, which is independent of observations, is
given by

$$\mathbf{L}_i(k \mid k-1) = \mathbf{Y}_i(k \mid k-1)\mathbf{F}(k)\mathbf{Y}_i^{-1}(k-1 \mid k-1). \tag{3.14}$$

## Estimation

The predictions are combined with information from local observations
to compute partial information estimates, $\tilde{\mathbf{y}}_j(k \mid k)$ and $\tilde{\mathbf{Y}}_j(k \mid k)$. These
are then communicated between all nodes in a fully connected network.
At each node, after communication, they are assimilated to produce global
information estimates.

$$\hat{\mathbf{y}}_i(k \mid k) = \hat{\mathbf{y}}_i(k \mid k-1) + \sum_{j=1}^{N}[\tilde{\mathbf{y}}_j(k \mid k) - \hat{\mathbf{y}}_i(k \mid k-1)] \tag{3.15}$$

$$\mathbf{Y}_i(k \mid k) = \mathbf{Y}_i(k \mid k-1) + \sum_{j=1}^{N}\left[\tilde{\mathbf{Y}}_j(k \mid k) - \mathbf{Y}_i(k \mid k-1)\right], \tag{3.16}$$

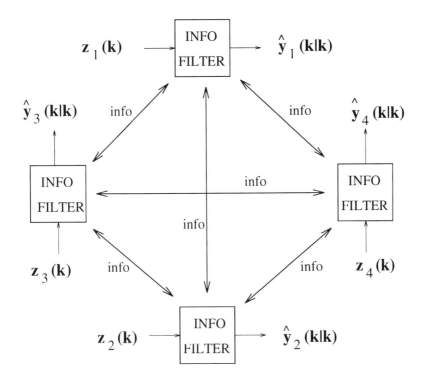

**FIGURE 3.7**
**Decentralized Information Filter**

where the local partial estimates are obtained from local information predictions and information from local observations,

$$\tilde{\mathbf{y}}_j(k \mid k) = \hat{\mathbf{y}}_j(k \mid k - 1) + \mathbf{i}_j(k) \tag{3.17}$$

$$\tilde{\mathbf{Y}}_j(k \mid k) = \mathbf{Y}_j(k \mid k - 1) + \mathbf{I}_j(k). \tag{3.18}$$

This is the decentralized Information filter (**DIF**) [48], [71], [87]. It has an Information filter running at each node, with communication between the nodes occurring before assimilation.

If each node begins with a common initial information state estimate (e.g., $\hat{\mathbf{y}}_j(0 \mid 0) = \mathbf{0}$ and $\mathbf{Y}_j(0 \mid 0) = \mathbf{0}$) and the network is fully connected, then the global estimates $\hat{\mathbf{y}}_i(k \mid k)$ and $\mathbf{Y}_i(k \mid k)$ obtained by each node are identical. This means that the predicted information vectors for any two nodes $i$ and $j$ are identical (from Equations 3.13 and 3.12), i.e.,

$$\hat{\mathbf{y}}_i(k \mid k - 1) = \hat{\mathbf{y}}_j(k \mid k - 1)$$
$$\mathbf{Y}_i(k \mid k - 1) = \mathbf{Y}_j(k \mid k - 1).$$

Invoking this equality in Equations 3.17 and 3.18 leads to

$$\tilde{\mathbf{y}}_j(k \mid k) - \hat{\mathbf{y}}_i(k \mid k - 1) = \mathbf{i}_j(k)$$
$$\tilde{\mathbf{Y}}_j(k \mid k) - \mathbf{Y}_i(k \mid k - 1) = \mathbf{I}_j(k).$$

These two equations are then used to simplify the assimilation Equations 3.15 and 3.16 to

$$\hat{\mathbf{y}}_i(k \mid k) = \hat{\mathbf{y}}_i(k \mid k - 1) + \sum_{j=1}^{N} \mathbf{i}_j(k) \qquad (3.19)$$

$$\mathbf{Y}_i(k \mid k) = \mathbf{Y}_i(k \mid k - 1) + \sum_{j=1}^{N} \mathbf{I}_j(k). \qquad (3.20)$$

This is the DIF with the same initial filter states at each node, in a fully connected network. The estimates obtained by each node are exactly the same for all nodes, and indeed are identical to those given by an equivalent centralized system [48], [87], [71]. Each node receives all information contributions from the other nodes and computes a global information estimate identical to that obtained by central fusion. Such a decentralized information filtering network is illustrated in Figure 3.7 for four fully connected nodes. The DIF algorithm can be summarized as consisting of the following local steps: information prediction, observation, internodal communication, assimilation and global information estimation.

It is important to note that for this algorithm to give exactly the same results as those of a corresponding centralized Information filter it must be run at *"full rate"*, which means that communication of information has to be carried out after every measurement [35], [36]. If the frequency of communication is less than the frequency of measurement, the DIF algorithm becomes suboptimal with respect to the centralized filter. With full communication rate, in addition to the benefits of decentralization, the global estimates obtained locally by the DIF are exactly the same as those produced by the centralized Information filter.

### 3.4.3   The Decentralized Kalman Filter (DKF)

For completeness, the decentralized Kalman filter (**DKF**), which is the state space version of the DIF is presented. The DKF algorithm is an algebraic equivalent of the DIF and was explicitly derived and implemented by Rao [106]. Its main use in this book is in Chapter 4 where it is used to demonstrate the advantages of information space internodal communication over state space communication.

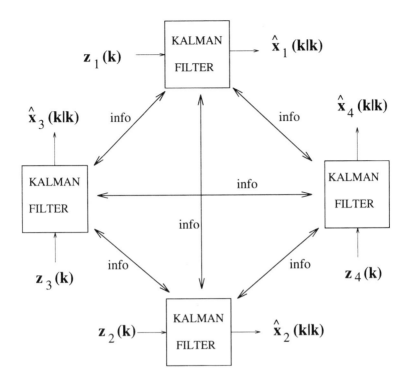

**FIGURE 3.8**
**Decentralized Kalman Filter**

## Prediction

The nodal state and variance prediction equations for the DKF directly follow from those of the Kalman filter. This is because each node carries out the conventional Kalman prediction steps. The system is fully connected, so the local models and vectors are the same as those in the centralized Kalman filter. Thus at each node $i$, the prediction equations are

$$\hat{\mathbf{x}}_i(k \mid k - 1) = \mathbf{F}(k)\hat{\mathbf{x}}_i(k - 1 \mid k - 1) + \mathbf{B}(k)\mathbf{u}_i(k - 1) \qquad (3.21)$$
$$\mathbf{P}_i(k \mid k - 1) = \mathbf{F}(k)\mathbf{P}_i(k - 1 \mid k - 1)\mathbf{F}^T(k) + \mathbf{Q}(k). \qquad (3.22)$$

Prediction is carried out locally before communication.

## Estimation

An outline of the DKF state and covariance estimation equations is presented. Details of their derivation are covered in [20] and [106]. What is of interest here is the nature of these equations. The update equations are

$$\hat{\mathbf{x}}_i(k \mid k) = \mathbf{P}_i(k \mid k) \left\{ \mathbf{P}_i^{-1}(k \mid k-1)\hat{\mathbf{x}}_i(k \mid k-1) + \sum_{j=1}^{N} \mathbf{e}_j(k) \right\}$$

$$\mathbf{P}_i(k \mid k) = \left[ \mathbf{P}_i^{-1}(k \mid k-1) + \sum_{j=1}^{N} \mathbf{E}_j(k) \right]^{-1},$$

where

$$\mathbf{e}_j(k) \triangleq \left\{ \tilde{\mathbf{P}}_j^{-1}(k \mid k)\tilde{\mathbf{x}}_j(k \mid k) - \mathbf{P}_j^{-1}(k \mid k-1)\hat{\mathbf{x}}_j(k \mid k-1) \right\} \qquad (3.23)$$

defines the *state error* information, with an associated *variance error* information given by

$$\mathbf{E}_j(k) \triangleq \left\{ \tilde{\mathbf{P}}_j^{-1}(k \mid k) - \mathbf{P}_j^{-1}(k \mid k-1) \right\}. \qquad (3.24)$$

$\tilde{\mathbf{P}}_j^{-1}(k \mid k)$ and $\tilde{\mathbf{x}}_j(k \mid k)$ are the local partial estimates based on local prediction and local observation at node $j$. They are computed according to local Kalman filter estimate equations:

$$\tilde{\mathbf{x}}_j(k \mid k) = \hat{\mathbf{x}}_j(k \mid k-1) + \mathbf{W}_j(k)\left[\mathbf{z}_j(k) - \mathbf{H}_j(k)\hat{\mathbf{x}}_j(k \mid k-1)\right] \qquad (3.25)$$

$$\tilde{\mathbf{P}}_j(k \mid k) = \mathbf{P}_j(k \mid k-1) - \mathbf{W}_j(k)\mathbf{S}_j(k)\mathbf{W}_j^T(k) \qquad (3.26)$$

where the local gain matrix is given by

$$\mathbf{W}_j(k) = \mathbf{P}_j(k \mid k-1)\mathbf{H}_j^T(k)\mathbf{S}_j^{-1}(k) \qquad (3.27)$$

and the local innovation covariance is

$$\mathbf{S}_j(k) = \mathbf{H}_j(k)\mathbf{P}_j(k \mid k-1)\mathbf{H}_j^T(k) + \mathbf{R}_j(k). \qquad (3.28)$$

The DKF algorithm is summarized in Figure 3.8 as a network consisting of communicating nodes with a Kalman filter running at each node. Although $\mathbf{e}_j(k)$ and $\mathbf{E}_j(k)$ are information variables, the data fusion process is essentially carried out in state space. Information is not formally defined and hence the benefits of information space are not best utilized.

### 3.4.4   The Decentralized Extended Information Filter (DEIF)

It is important to extend the decentralized estimation algorithms to deal with problems in nonlinear systems. This is done by decentralizing the extended Information filter algorithm to produce the decentralized extended

Information filter (**DEIF**). Its derivation follows from that of the DIF.

**Prediction**

Local information prediction takes place as in the DIF, but as in the EIF, nonlinear models and linearized models are used. Predictions are expressed as functions of these models and previous, locally determined, global information estimates,

$$\hat{\mathbf{y}}_i(k \mid k-1) = \mathbf{Y}_i(k \mid k-1)\mathbf{f}(k, \hat{\mathbf{x}}_i(k-1 \mid k-1), \mathbf{u}_i(k-1)) \qquad (3.29)$$

$$\mathbf{Y}_i(k \mid k-1) = \left[ \nabla \mathbf{f}_{x_i}(k)\mathbf{Y}_i^{-1}(k-1 \mid k-1)\nabla \mathbf{f}_{x_i}{}^T(k) + \mathbf{Q}(k) \right]^{-1}. \qquad (3.30)$$

The local nonlinear state transition and process noise models are the same as in the global system. Jacobians for the state transition model have a similar property since they are evaluated at the global, locally obtained, state predictions. These state predictions are the same for all nodes in fully connected topologies.

**Estimation**

The estimation and assimilation pattern is the same as for the DIF. However, the communicated information depends on linearized models and nonlinear observations. These are communicated in a fully connected topology and assimilated as follows:

$$\hat{\mathbf{y}}_i(k \mid k) = \hat{\mathbf{y}}_i(k \mid k-1) + \sum_{j=1}^{N} \mathbf{i}_j(k) \qquad (3.31)$$

$$\mathbf{Y}_i(k \mid k) = \mathbf{Y}_i(k \mid k-1) + \sum_{j=1}^{N} \mathbf{I}_j(k), \qquad (3.32)$$

where the associated information matrix and state information contribution from local observations are, respectively, given by

$$\mathbf{I}_j(k) = \nabla \mathbf{h}_{x_j}{}^T(k)\mathbf{R}_j^{-1}(k)\nabla \mathbf{h}_{x_j}(k) \qquad (3.33)$$

$$\mathbf{i}_j(k) = \nabla \mathbf{h}_{x_j}{}^T(k)\mathbf{R}_j^{-1}(k)[\nu_j(k) + \nabla \mathbf{h}_{x_j}(k)\hat{\mathbf{x}}_j(k \mid k-1)]. \qquad (3.34)$$

The vector $\nu_j(k)$ represents the local innovations,

$$\nu_j(k) = \mathbf{z}_j(k) - \mathbf{h}_j(\hat{\mathbf{x}}_j(k \mid k-1)). \qquad (3.35)$$

This completes an outline of the DEIF. This estimation algorithm is a novel result which constitutes a significant contribution to multisensor fusion. It provides an estimation algorithm for nonlinear multisensor systems by using variables expressed in terms of information. It is easy to initialize, which is a crucial property where linearized models are employed. However, the DEIF retains some of the drawbacks of the EIF. In particular, it is prone to linearization instabilities.

### 3.4.5   The Decentralized Extended Kalman Filter (DEKF)

Decentralized estimation for nonlinear systems is completed by presenting the state space algebraic equivalent of the DEIF. The extended Kalman filter has been introduced as a linearized estimator for nonlinear systems. It is obtained by a series expansion of the nonlinear dynamics and of the measurement equations. The EKF can be decentralized in the same fashion as the Kalman filter (in the DKF) to produce the decentralized extended Kalman filter (**DEKF**).

**Prediction**

The prediction equations of the DEKF follow from those of the EKF by localizing computation in the same way as is done for the DKF. Hence, prediction depends on local, nonlinear and linearized system models,

$$\hat{\mathbf{x}}_i(k \mid k-1) = \mathbf{f}\left(k, \hat{\mathbf{x}}_i(k-1 \mid k-1), \mathbf{u}_i(k-1)\right) \tag{3.36}$$

$$\mathbf{P}_i(k \mid k-1) = \nabla \mathbf{f}_{x_i}(k)\mathbf{P}_i(k-1 \mid k-1)\nabla \mathbf{f}_{x_i}{}^T(k) + \mathbf{Q}(k). \tag{3.37}$$

The system is assumed fully connected and local nonlinear state transition and noise models are the same as in the global system. The state transition Jacobians, although locally computed, are the same for all nodes as they are evaluated at the predicted global state, which is the same for all nodes.

**Estimation**

Local predictions are used to compute linearized, local partial estimates which are then communicated between sensor nodes and locally assimilated to give global estimates for the nonlinear system. The estimation and assimilation equations are the same as in the DKF. What is different is the way the communicated partial estimates are computed from local nonlinear observations and linearized local models. The entire estimation process is presented here for completeness and to clarify the computational stages.

$$\hat{\mathbf{x}}_i(k \mid k) = \mathbf{P}_i(k \mid k)\left\{\mathbf{P}_i^{-1}(k \mid k-1)\hat{\mathbf{x}}_i(k \mid k-1) + \sum_{j=1}^{N}\mathbf{e}_j(k)\right\}$$

$$\mathbf{P}_i(k \mid k) = \left[\mathbf{P}_i^{-1}(k \mid k-1) + \sum_{j=1}^{N}\mathbf{E}_j(k)\right]^{-1},$$

where, as for the DKF,

$$\mathbf{e}_j(k) \triangleq \left\{\tilde{\mathbf{P}}_j^{-1}(k \mid k)\tilde{\mathbf{x}}_j(k \mid k) - \mathbf{P}_j^{-1}(k \mid k-1)\hat{\mathbf{x}}_j(k \mid k-1)\right\},$$

defines the state error information, with an associated variance error

$$\mathbf{E}_j(k) \triangleq \left\{\tilde{\mathbf{P}}_j^{-1}(k \mid k) - \mathbf{P}_j^{-1}(k \mid k-1)\right\}. \tag{3.38}$$

$\tilde{\mathbf{P}}_j^{-1}(k \mid k)$ and $\tilde{\mathbf{x}}_j(k \mid k)$ are the local partial estimates based on local prediction and local nonlinear observation at node $j$. They are computed according to local EKF estimate equations:

$$\tilde{\mathbf{x}}_j(k \mid k) = \hat{\mathbf{x}}_j(k \mid k - 1) + \mathbf{W}_j(k)\left[\mathbf{z}_j(k) - \mathbf{h}_j(\hat{\mathbf{x}}_j(k \mid k - 1))\right] \quad (3.39)$$

$$\tilde{\mathbf{P}}_j(k \mid k) = \mathbf{P}_j(k \mid k - 1) - \mathbf{W}_j(k)\mathbf{S}_j(k)\mathbf{W}_j^T(k). \quad (3.40)$$

The local linearized gain $\mathbf{W}_j(k)$ is given as

$$\mathbf{W}_j(k) = \mathbf{P}_j(k \mid k - 1)\nabla \mathbf{h}_{x_j}{}^T(k)\mathbf{S}_j^{-1}(k) \quad (3.41)$$

and the local linearized innovation variance is

$$\mathbf{S}_j(k) = \nabla \mathbf{h}_{x_j}(k)\mathbf{P}_j(k \mid k - 1)\nabla \mathbf{h}_{x_j}{}^T(k) + \mathbf{R}_j(k). \quad (3.42)$$

Like the DKF, the DEKF will be used in Chapter 4 to distinguish between information and state space internodal communication.

## 3.5 The Limitations of Fully Connected Decentralization

The four decentralized filter algorithms discussed above, the DKF, DIF, DEKF and DEIF allow decentralized estimation to be carried out. Thus, the estimation problem involving an arbitrary number of communicating sensor nodes for both linear and nonlinear systems has been solved. By use of local observations and communicated information, each node is capable of independent local estimation without the need for a global observer or central processor. However, it is important to emphasize that for locally obtained global estimates to be the same as those obtained in equivalent centralized systems, the sensor nodes must be fully connected, and a global model must be maintained by all sensing nodes.

Figure 3.9 shows a fully connected network of nine nodes. As the number of nodes increases, severe difficulties arise.

- **Communication:** In a fully connected system there is excessive redundant communication. Nodes that need not communicate do communicate. For those that need to communicate, information that does not need to be exchanged is exchanged. Consequently, there is wastage both in terms of number of communication links and size of communicated messages. In many applications the communication requirements of a fully connected topology are difficult to meet.

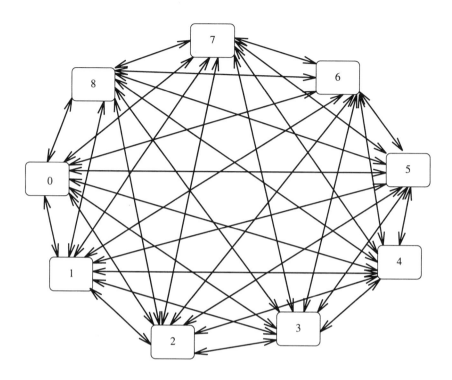

**FIGURE 3.9**
**Fully Connected Network (the Limitations)**

- **Computation:** The local models, information and state matrices are
  of the same size as those in an equivalent centralized system. As a
  result there is no significant reduction of memory requirements and
  computational load. There is a lot of redundant processing. This
  redundancy does not improve survivability; it only wastes computational resources.

- **Hardware Constraints:** The assumption of a fully connected topology is, in general, unrealistic. This assumption is invalidated by loss
  of one or more communication links. As the number of nodes $N$ increases, the number of physical links required by each node, $N - 1$,
  and the overall number of links, $N(N-1)/2$, increase. Such extensive
  physical connections are not easily supported in hardware.

These problems affect fully connected decentralized estimation systems and
in this way undermine their practical usefulness. Resolving these limitations
creates the motivation for the work described in Chapter 4.

## 3.6 Summary

This chapter has introduced multisensor systems, the data fusion problem, data fusion methods and architectures. Multisensor data fusion methods were briefly surveyed and discussed. The concept of complete decentralization was defined and the advantages of decentralized systems discussed. This chapter effectively extended the single sensor estimation algorithms of Chapter 2 to multisensor systems. This was done by decentralizing the observer first and explaining why the conventional form of the Kalman filter cannot be used. Decentralized estimation algorithms were then derived by using the decentralized observer and the information form of the Kalman filter.

As a result, four fully connected decentralized estimation algorithms were presented: the DKF, DIF, DEKF and DEIF. In this way, fully decentralized sensing was shown to be algorithmically attainable. Information space is the logical place to decentralize, communicate and assimilate information variables. The DEIF is more practically useful than the DIF. It is the key contribution of this chapter. For the decentralized estimation algorithms to give exactly the same results as corresponding centralized algorithms they must be run at full communication rate. The main drawback of the four estimation algorithms is the requirement for a fully connected topology. Problems arising from this constraint are outlined, creating the motivation for the material presented in Chapter 4.

# Chapter 4

## Scalable Decentralized Estimation

### 4.1 Introduction

This chapter seeks to resolve the problems associated with fully connected decentralized estimation: limited scalability, redundant communication, excessive computation and vulnerability to communication link loss. Two related techniques are proposed as a composite solution: model distribution and model defined internodal communication. Model distribution reduces the size of local models by involving only locally relevant states at each node. Internodal local communication ensures that only nodes that have to communicate do so, and that when they do, they only exchange relevant information.

The chapter starts by introducing the principles of model distribution and proceeds to present a new technique for deriving nodal transformation matrices. The main principle of this technique lies in the identification of the states (information) which need to be known locally and then distributing global models on this basis. Since, in general, transformation matrices are not invertible, the Moore-Penrose generalized inverse is introduced. Its relevant special properties and theorems are outlined. Distribution of models is then discussed. A general internodal transformation theory is then developed for both state space and information space. This serves to formalize the communication and network topology requirements by identifying nodes that have to communicate and the information they need to share.

Special cases of the general transformation are presented and their possible applications explored. Internodal communication theory and model distribution are then applied to the decentralized estimation algorithms of Chapter 3. This produces scalable decentralized estimation which is free of the drawbacks of fully connected systems. The resulting four distributed and decentralized filters are compared, contrasted and evaluated.

### 4.1.1   Model Distribution

Model distribution is the process of constructing reduced order models from a global system model by creating local state vectors which consist of locally relevant states such that there is dynamic equivalence between local and global models. Model distribution can be employed in decentralized estimation to provide a system that does not require fully connected topologies. The general distribution approach is due to [21], although the techniques and results presented here are much more general and exhaustive than those in [21] and [22].

The nodal state vector $\mathbf{x}_j(k)$ is related to the global state vector $\mathbf{x}(k)$ by

$$\mathbf{x}_j(k) = \mathbf{T}_j(k)\mathbf{x}(k), \qquad (4.1)$$

where $\mathbf{T}_j(k)$ is a linear *nodal transformation matrix*. As expressed in Equation 3.5, the local observations are obtained from the global state vector according to

$$\mathbf{z}_j(k) = \mathbf{H}_j(k)\mathbf{x}(k) + \mathbf{v}_j(k). \qquad (4.2)$$

This local observation vector can also be obtained from the local state vector through a local model in the form

$$\mathbf{z}_j(k) = \mathbf{C}_j(k)\mathbf{x}_j(k) + \mathbf{v}_j(k), \qquad (4.3)$$

where $\mathbf{C}_j(k)$ is the distributed local observation model. The partitioned observation model $\mathbf{H}_j(k)$ is then related to the local observation matrix $\mathbf{C}_j(k)$ by equating Equations 4.2 and 4.3.

$$\mathbf{H}_j(k)\mathbf{x}(k) = \mathbf{C}_j(k)\mathbf{x}_j(k) \qquad (4.4)$$

Substituting Equation 4.1 into Equation 4.4 gives

$$\begin{aligned} \mathbf{H}_j(k)\mathbf{x}(k) &= \mathbf{C}_j(k)\mathbf{T}_j(k)\mathbf{x}(k) \\ \Leftrightarrow \mathbf{H}_j(k) &= \mathbf{C}_j(k)\mathbf{T}_j(k). \end{aligned} \qquad (4.5)$$

This is a sufficient but not necessary condition on the two observation models. The internodal communication theory developed in Section 4.4 utilizes this property to minimize communication.

### 4.1.2   Nodal Transformation Determination

The next issue is the determination of $\mathbf{T}_j(k)$. One approach would be to define $\mathbf{T}_j(k)$ as any matrix that *arbitrarily* picks states or combinations of states from the global state vector $\mathbf{x}(k)$. However, such an approach would present problems when dealing with communication and assimilation of information between nodes. There will be a need to propagate information between two *unconnected* nodes which have a common state space observed by either or both nodes. This leads to the problem of integrating information common to two communicating nodes.

The problem could be solved by an additional Information filter, the channel filter, as proposed in the non-fully connected topologies discussed in [48] and [38]. However, the additional filtering process at each node increases computational load. The question then becomes how to derive nodal transformation matrices such that the local states can be locally estimated and controlled as optimally as they would be centrally *without* having to propagate information between unconnected nodes. The idea is to take a general system and determine how its models and state vector can be distributed while satisfying this constraint. In doing this the main principle is to identify what needs to be known locally and then distributing the global models on that basis. Consequently, $\mathbf{T}_j(k)$ picks states or combinations of states from the global vector to form a reduced order local state. With this approach, the need for channel filters does not arise.

## 4.2 An Extended Example

Consider the mass system depicted in Figure 4.1. This system consists of four trolley masses interconnected by springs. Input in any one affects the other three. It is chosen because it is the simplest example possessing all the characteristic properties of an interconnected, multisensor, multiactuator dynamic system. It is used here to show how nodal transformations can be derived by using the system model $\mathbf{F}(k)$ to identify those states that are locally relevant.

First, the case where the local states are unscaled, locally relevant, individual global states is developed. The case of local states proportionally dependent on individual global states is then considered. Finally, the case of linear combinations of global states as local states is developed. The results are then generalized to cover any coupled system.

### 4.2.1 Unscaled Individual States

The position and velocity of a general mass $j$ are denoted by

$$\mathbf{x}_j^m(k) = \begin{bmatrix} x_j \\ \dot{x}_j \end{bmatrix}. \tag{4.6}$$

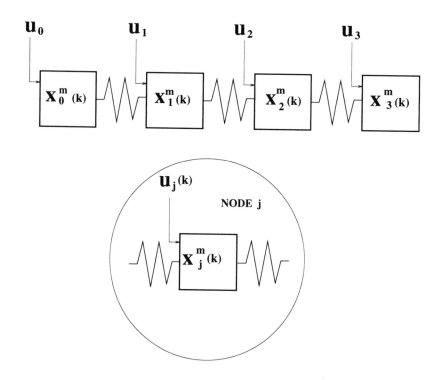

**FIGURE 4.1**
**Coupled Mass System**

The global (central) state vector $\mathbf{x}(k)$ which consists of all the states of the system can be defined as follows:

$$
\begin{bmatrix} \mathbf{x}_0^m(k) \\ \mathbf{x}_1^m(k) \\ \mathbf{x}_2^m(k) \\ \mathbf{x}_3^m(k) \end{bmatrix} = \begin{bmatrix} x_0 \\ \dot{x}_0 \\ x_1 \\ \dot{x}_1 \\ x_2 \\ \dot{x}_2 \\ x_3 \\ \dot{x}_3 \end{bmatrix}. \tag{4.7}
$$

Rearranging the states so that the positions and velocities are grouped together gives the global state vector

$$\mathbf{x}(k) = \begin{bmatrix} x_0 \\ x_1 \\ x_2 \\ x_3 \\ x_4 \\ x_5 \\ x_6 \\ x_7 \end{bmatrix}, \tag{4.8}$$

where the last four states are the velocities, i.e.,

$$\begin{bmatrix} x_4 \\ x_5 \\ x_6 \\ x_7 \end{bmatrix} = \begin{bmatrix} \dot{x}_0 \\ \dot{x}_1 \\ \dot{x}_2 \\ \dot{x}_3 \end{bmatrix}. \tag{4.9}$$

Consider a general node $j$ which estimates and controls the position and velocity of mass $j$. From the state transition matrix of *any* chain mass system of the form shown in Figure 4.1, the state vector $\mathbf{x}_j^m(k)$ and control vector $\mathbf{u}_j(k)$ directly depend on state vectors $\mathbf{x}_{j-1}^m(k-1)$, $\mathbf{x}_j^m(k-1)$ and $\mathbf{x}_{j+1}^m(k-1)$. For example, for the structure in Figure 4.1 the vectors $\mathbf{x}_1^m(k)$ and $\mathbf{u}_1(k)$ directly depend on the vectors $\mathbf{x}_0^m(k-1)$, $\mathbf{x}_1^m(k-1)$ and $\mathbf{x}_2^m(k-1)$. This then determines the *local* state vector $\mathbf{x}_j(k)$ which contains all the states necessary at node $j$ in order to effectively estimate and control the mass at $j$ locally. As in the case of the global vector the positions and velocities are grouped together. The local state models are thus

$$\mathbf{x}_0(k) = \begin{bmatrix} x_0 \\ x_1 \\ x_4 \\ x_5 \end{bmatrix}, \qquad \mathbf{x}_1(k) = \begin{bmatrix} x_0 \\ x_1 \\ x_2 \\ x_4 \\ x_5 \\ x_6 \end{bmatrix}, \tag{4.10}$$

$$\mathbf{x}_2(k) = \begin{bmatrix} x_1 \\ x_2 \\ x_3 \\ x_5 \\ x_6 \\ x_7 \end{bmatrix}, \qquad \mathbf{x}_3(k) = \begin{bmatrix} x_2 \\ x_3 \\ x_6 \\ x_7 \end{bmatrix}. \tag{4.11}$$

In this way the global state vector, $\mathbf{x}(k)$, is distributed into local vectors, $\mathbf{x}_j(k)$, containing only locally relevant states. The corresponding nodal transformation matrices are obtained from Equation 4.1.

$$\mathbf{T}_0(k) = \begin{bmatrix} 1 & 0 & 0 & 0 & 0 & 0 & 0 & 0 \\ 0 & 1 & 0 & 0 & 0 & 0 & 0 & 0 \\ 0 & 0 & 0 & 0 & 1 & 0 & 0 & 0 \\ 0 & 0 & 0 & 0 & 0 & 1 & 0 & 0 \end{bmatrix} \tag{4.12}$$

$$\mathbf{T}_1(k) = \begin{bmatrix} 1 & 0 & 0 & 0 & 0 & 0 & 0 & 0 \\ 0 & 1 & 0 & 0 & 0 & 0 & 0 & 0 \\ 0 & 0 & 1 & 0 & 0 & 0 & 0 & 0 \\ 0 & 0 & 0 & 0 & 1 & 0 & 0 & 0 \\ 0 & 0 & 0 & 0 & 0 & 1 & 0 & 0 \\ 0 & 0 & 0 & 0 & 0 & 0 & 1 & 0 \end{bmatrix} \tag{4.13}$$

$$\mathbf{T}_2(k) = \begin{bmatrix} 0 & 1 & 0 & 0 & 0 & 0 & 0 & 0 \\ 0 & 0 & 1 & 0 & 0 & 0 & 0 & 0 \\ 0 & 0 & 0 & 1 & 0 & 0 & 0 & 0 \\ 0 & 0 & 0 & 0 & 0 & 1 & 0 & 0 \\ 0 & 0 & 0 & 0 & 0 & 0 & 1 & 0 \\ 0 & 0 & 0 & 0 & 0 & 0 & 0 & 1 \end{bmatrix} \tag{4.14}$$

$$\mathbf{T}_3(k) = \begin{bmatrix} 0 & 0 & 1 & 0 & 0 & 0 & 0 & 0 \\ 0 & 0 & 0 & 1 & 0 & 0 & 0 & 0 \\ 0 & 0 & 0 & 0 & 0 & 0 & 1 & 0 \\ 0 & 0 & 0 & 0 & 0 & 0 & 0 & 1 \end{bmatrix} . \tag{4.15}$$

Thus, given a physical system the nodal transformation matrix $\mathbf{T}_j(k)$ is dependent on the way states physically influence each other, that is, $\mathbf{T}_j(k)$ is dependent on the state transition matrix, $\mathbf{F}(k)$. In general, the elements of $\mathbf{T}_j(k)$ are not necessarily *ones* and *zeros*. This is demonstrated in the following sections.

## 4.2.2    Proportionally Dependent States

The nodal transformation matrices given above for the mass system are not unique, and, indeed, the local models of this illustrative example can be generalized. Let the global state transition matrix of the system be represented as follows:

$$\mathbf{F}(k) = \begin{bmatrix} f_{00} & 0 & 0 & 0 & f_{04} & 0 & 0 & 0 \\ 0 & f_{11} & 0 & 0 & 0 & f_{15} & 0 & 0 \\ 0 & 0 & f_{22} & 0 & 0 & 0 & f_{26} & 0 \\ 0 & 0 & 0 & f_{33} & 0 & 0 & 0 & f_{37} \\ f_{40} & f_{41} & 0 & 0 & f_{44} & 0 & 0 & 0 \\ f_{50} & f_{51} & f_{52} & 0 & 0 & f_{55} & 0 & 0 \\ 0 & f_{61} & f_{62} & f_{63} & 0 & 0 & f_{66} & 0 \\ 0 & 0 & f_{72} & f_{73} & 0 & 0 & 0 & f_{77} \end{bmatrix}, \tag{4.16}$$

where this matrix follows from interconnection of the masses as shown in Figure 4.1. Let the states to be estimated and controlled by node 1 be proportional to the velocity and position of mass 1 such that

$$\mathbf{x}_1^l(k) = \mathbf{A}_1 \mathbf{x}_1^m(k) \tag{4.17}$$

$$\tag{4.18}$$

$$= \begin{bmatrix} a_{01} & 0 \\ 0 & a_{05} \end{bmatrix} \begin{bmatrix} x_1(k) \\ x_5(k) \end{bmatrix} \tag{4.19}$$

$$\tag{4.20}$$

$$= \begin{bmatrix} a_{01} x_1(k) \\ a_{05} x_5(k) \end{bmatrix}, \tag{4.21}$$

where $\mathbf{A}_1$ is a diagonal matrix with general non-zero constant elements $a_{01}$ and $a_{05}$. From the state transition matrix it follows that

$\mathbf{x}_1^l(k)$

$$= \begin{bmatrix} a_{01}\{f_{11}x_1(k-1) + f_{15}x_5(k-1)\} \\ a_{02}\{f_{50}x_0(k-1) + f_{51}x_1(k-1) + f_{52}x_2(k-1) + f_{55}x_5(k-1)\} \end{bmatrix}$$

$$= \begin{bmatrix} 0 & a_{01}f_{11} & 0 & a_{01}f_{15} \\ a_{02}f_{50} & a_{02}f_{51} & a_{02}f_{52} & a_{02}f_{55} \end{bmatrix} \begin{bmatrix} x_0(k-1) \\ x_1(k-1) \\ x_2(k-1) \\ x_5(k-1) \end{bmatrix}.$$

Thus the general, minimum sized, local state vector, $\mathbf{x}_1(k)$, required at node 1 in order to effectively estimate and control the states in $\mathbf{x}_1^l(k)$ locally is given by

$$\mathbf{x}_1(k) = \begin{bmatrix} \alpha_{00} x_0(k) \\ \alpha_{11} x_1(k) \\ \alpha_{22} x_2(k) \\ \alpha_{35} x_5(k) \end{bmatrix}, \tag{4.22}$$

where $f_{ij}$, an element of $\mathbf{F}(k)$, operates as an indicator showing relevant states and $\alpha_{ij}$ is a strictly non-zero arbitrary constant. Comparing this vector with the one in Equation 4.10 shows that the vector in Equation 4.22 is of lower order. The model in Equation 4.22 demonstrates that the velocities of masses 0 and 2 are *strictly* not required to determine the position and velocity of mass 1, only their positions are essential. Thus, the issue of which states are essential to estimate and control any given states is determined by $\mathbf{F}(k)$. Equation 4.1 shows that

$$\mathbf{T}_1(k) = \begin{bmatrix} \alpha_{00} & 0 & 0 & 0 & 0 & 0 & 0 \\ 0 & \alpha_{11} & 0 & 0 & 0 & 0 & 0 \\ 0 & 0 & \alpha_{22} & 0 & 0 & 0 & 0 \\ 0 & 0 & 0 & 0 & 0 & \alpha_{35} & 0 & 0 \end{bmatrix}, \tag{4.23}$$

is the general nodal transformation matrix required by node 1 for tracking and controlling local states.

### 4.2.3  Linear Combination of States

A further case of interest is when the local states are linear combinations of global states. For example, using the mass system, let the state vector of interest at node 1 be a linear combination of positions and velocities of masses 0 and 1, i.e.,

$$
\mathbf{x}_1^l(k) = \begin{bmatrix} x_0^l(k) \\ x_4^l(k) \end{bmatrix}
$$

$$
= \mathbf{A}_0 \mathbf{x}_0^m(k) + \mathbf{A}_1 \mathbf{x}_1^m(k)
$$

$$
= \begin{bmatrix} a_{00} & 0 \\ 0 & a_{04} \end{bmatrix} \begin{bmatrix} x_0(k) \\ x_4(k) \end{bmatrix} + \begin{bmatrix} a_{01} & 0 \\ 0 & a_{05} \end{bmatrix} \begin{bmatrix} x_1(k) \\ x_5(k) \end{bmatrix}
$$

$$
= \begin{bmatrix} a_{00} x_0(k) + a_{01} x_1(k) \\ a_{04} x_4(k) + a_{05} x_5(k) \end{bmatrix},
$$

where $a_{00}$, $a_{04}$, $a_{01}$ and $a_{05}$ are general non-zero constants.

From the state transition matrix it follows that

$$
\mathbf{x}_1^l(k) = \begin{bmatrix} a_{00}\{f_{00}x_0(k-1) + f_{04}x_4(k-1)\} + a_{01}\{f_{11}x_1(k-1) \\ +f_{15}x_5(k-1)\} \\ \\ a_{04}\{f_{40}x_0(k-1) + f_{41}x_1(k-1) + f_{44}x_4(k-1)\} \\ +a_{05}\{f_{50}x_0(k-1) + f_{51}x_1(k-1) + f_{52}x_2(k-1) \\ +f_{55}x_5(k-1)\} \end{bmatrix}
$$

$$
= \begin{bmatrix} a_{00}\{f_{00}x_0(k-1) + f_{04}x_4(k-1)\} + a_{01}\{f_{11}x_1(k-1) \\ +f_{15}x_5(k-1)\} \\ \\ [a_{04}f_{40} + a_{05}f_{50}]x_0(k-1) + [a_{04}f_{41} + a_{05}f_{51}]x_1(k-1) \\ +[a_{05}f_{52}]x_2(k-1) + [a_{05}f_{44}]x_4(k-1) + [a_{05}f_{55}]x_5(k-1) \end{bmatrix}
$$

$$
= \begin{bmatrix} a_{00}f_{00} & a_{01}f_{11} & 0 & a_{00}f_{04} & a_{01}f_{15} \\ \\ \{a_{04}f_{40} + a_{05}f_{50}\} & \{a_{04}f_{41} + a_{05}f_{51}\} & a_{05}f_{52} & a_{05}f_{44} & a_{05}f_{55} \end{bmatrix}
$$

$$\times \begin{bmatrix} x_0(k-1) \\ x_1(k-1) \\ x_2(k-1) \\ x_4(k-1) \\ x_5(k-1) \end{bmatrix}.$$

Therefore, the local state vector $\mathbf{x}_1(k)$ required at node 1 in order to effectively estimate and control the linear combinations of states in $\mathbf{x}_1^l(k)$ locally is given by

$$\mathbf{x}_1(k) = \begin{bmatrix} \alpha_{00} x_0(k) \\ \alpha_{11} x_1(k) \\ \alpha_{22} x_2(k) \\ \alpha_{34} x_4(k) \\ \alpha_{45} x_5(k) \end{bmatrix}, \qquad (4.24)$$

where $f_{ij}$, an element of $\mathbf{F}(k)$, operates as an indicator showing relevant states and $\alpha_{ij}$s are strictly non-zero arbitrary constants. Considering the model distribution result in Equation 4.1 this gives the general nodal transformation matrix $\mathbf{T}_1(k)$ required for a state vector of linear combination of states, $\mathbf{x}_1^l(k)$, as

$$\mathbf{T}_1(k) = \begin{bmatrix} \alpha_{00} & 0 & 0 & 0 & 0 & 0 & 0 & 0 \\ 0 & \alpha_{11} & 0 & 0 & 0 & 0 & 0 & 0 \\ 0 & 0 & \alpha_{22} & 0 & 0 & 0 & 0 & 0 \\ 0 & 0 & 0 & 0 & \alpha_{34} & 0 & 0 & 0 \\ 0 & 0 & 0 & 0 & 0 & \alpha_{45} & 0 & 0 \end{bmatrix}.$$

Although the preceding example addresses the issue of linear combination of states, the resulting local state vector $\mathbf{x}_1(k)$ given in Equation 4.24 does not have any linear combination of states as local states. It is possible to have this arrangement without creating problems with redundant states. A nodal transformation creating linear combination of states as local states does not give rise to redundant states, unless there are such states in the original system. This is because this derived transformation changes the basis and model size, but not the dynamic properties of the system. Consider the preceding example. From the expression for the current state, the previous state vector is given by

$$\mathbf{x}_1^l(k-1) = \mathbf{A}_0 \mathbf{x}_0^m(k-1) + \mathbf{A}_1 \mathbf{x}_1^m(k-1)$$

$$= \mathbf{A}_0 \begin{bmatrix} x_0(k-1) \\ x_4(k-1) \end{bmatrix} + \mathbf{A}_1 \begin{bmatrix} x_1(k-1) \\ x_5(k-1) \end{bmatrix}$$

$$\Leftrightarrow \begin{bmatrix} x_0(k-1) \\ x_4(k-1) \end{bmatrix} = \mathbf{A}_0^{-1} \begin{bmatrix} x_0^l(k-1) \\ x_4^l(k-1) \end{bmatrix} - \mathbf{A}_0^{-1} \mathbf{A}_1 \begin{bmatrix} x_1(k-1) \\ x_5(k-1) \end{bmatrix}$$

$$
= \begin{bmatrix} \frac{1}{a_{00}} & 0 \\ 0 & \frac{1}{a_{04}} \end{bmatrix} \begin{bmatrix} x_0^l(k-1) \\ x_4^l(k-1) \end{bmatrix} - \begin{bmatrix} \frac{a_{01}}{a_{00}} & 0 \\ 0 & \frac{a_{05}}{a_{04}} \end{bmatrix} \begin{bmatrix} x_1(k-1) \\ x_5(k-1) \end{bmatrix}
$$

$$
= \begin{bmatrix} \frac{1}{a_{00}} x_0^l(k-1) - \frac{a_{01}}{a_{00}} x_1(k-1) \\ \frac{1}{a_{04}} x_4^l(k-1) - \frac{a_{05}}{a_{04}} x_5(k-1) \end{bmatrix}.
$$

These expressions of $x_0(k-1)$ and $x_4(k-1)$ can be substituted into the equation for $\mathbf{x}_1^l(k)$,

$$
\mathbf{x}_1^l(k)
$$

$$
= \begin{bmatrix} a_{00}f_{00} & a_{01}f_{11} & 0 & a_{00}f_{04} & a_{01}f_{15} \\ \{a_{04}f_{40} + a_{05}f_{50}\} & \{a_{04}f_{41} + a_{05}f_{51}\} & a_{05}f_{52} & a_{05}f_{44} & a_{05}f_{55} \end{bmatrix} \times
$$

$$
\begin{bmatrix} \frac{1}{a_{00}} x_0^l(k-1) - \frac{a_{01}}{a_{00}} x_1(k-1) \\ x_1(k-1) \\ x_2(k-1) \\ \frac{1}{a_{04}} x_4^l(k-1) - \frac{a_{05}}{a_{04}} x_5(k-1) \\ x_5(k-1) \end{bmatrix}
$$

$$
= \begin{bmatrix} 1 & a_{01}\{f_{11} - f_{00}\} & 0 & \frac{a_{00}}{a_{04}}f_{04} & \{a_{01}f_{15} - a_{00}f_{04}\frac{a_{05}}{a_{04}}\} \\ \{\frac{a_{04}}{a_{00}}f_{40} + \frac{a_{05}}{a_{00}}f_{50}\} & e_{11} & a_{05}f_{52} & \frac{a_{05}}{a_{04}}f_{44} & a_{05}\{f_{55} - \frac{a_{05}}{a_{04}}f_{44}\} \end{bmatrix}
$$

$$
\times \begin{bmatrix} x_0^l(k-1) \\ x_1(k-1) \\ x_2(k-1) \\ x_4^l(k-1) \\ x_5(k-1) \end{bmatrix}
$$

(*where* $e_{11} = a_{04}f_{41} + a_{05}f_{51} - \dfrac{a_{01}}{a_{00}}(a_{04}f_{40} + a_{05}f_{50})$ )

$$
= \begin{bmatrix} 1 & a_{01}\{f_{11} - f_{00}\} & 0 & \frac{a_{00}}{a_{04}}f_{04} & \{a_{01}f_{15} - a_{00}f_{04}\frac{a_{05}}{a_{04}}\} \\ \{\frac{a_{04}}{a_{00}}f_{40} + \frac{a_{05}}{a_{00}}f_{50}\} & e_{11} & a_{05}f_{52} & \frac{a_{05}}{a_{04}}f_{44} & a_{05}\{f_{55} - \frac{a_{05}}{a_{04}}f_{44}\} \end{bmatrix}
$$

$$\times \begin{bmatrix} a_{00}x_0(k) + a_{01}x_1(k) \\ \\ x_1(k-1) \\ \\ x_2(k-1) \\ \\ a_{04}x_4(k) + a_{05}x_5(k) \\ \\ x_5(k-1) \end{bmatrix}$$

Therefore, the general state vector $\mathbf{x}_1(k)$ with linear combination of states as local states becomes

$$\mathbf{x}_1(k) = \begin{bmatrix} \alpha_{00}x_0(k) + \alpha_{01}x_1(k) \\ \\ \alpha_{11}x_1(k) \\ \\ \alpha_{22}x_2(k) \\ \\ \alpha_{34}x_4(k) + \alpha_{35}x_5(k) \\ \\ \alpha_{45}x_5(k) \end{bmatrix}, \qquad (4.25)$$

where $f_{ij}$, an element of $\mathbf{F}(k)$, operates as an indicator showing relevant states and $\alpha_{ij}$s are strictly non-zero arbitrary constants. Using the model distribution result in Equation 4.1 gives the general nodal transformation matrix required for a state vector of linear combination of states $\mathbf{x}_1^l(k)$.

$$\mathbf{T}_1^l(k) = \begin{bmatrix} \alpha_{00} & \alpha_{01} & 0 & 0 & 0 & 0 & 0 & 0 \\ 0 & \alpha_{11} & 0 & 0 & 0 & 0 & 0 & 0 \\ 0 & 0 & \alpha_{22} & 0 & 0 & 0 & 0 & 0 \\ 0 & 0 & 0 & 0 & \alpha_{34} & \alpha_{35} & 0 & 0 \\ 0 & 0 & 0 & 0 & 0 & \alpha_{45} & 0 & 0 \end{bmatrix}$$

$$= \begin{bmatrix} 1 & \frac{\alpha_{01}}{\alpha_{11}} & 0 & 0 & 0 \\ 0 & 1 & 0 & 0 & 0 \\ 0 & 0 & 1 & 0 & 0 \\ 0 & 0 & 0 & 1 & \frac{\alpha_{35}}{\alpha_{45}} \\ 0 & 0 & 0 & 0 & 1 \end{bmatrix} \begin{bmatrix} \alpha_{00} & 0 & 0 & 0 & 0 & 0 & 0 & 0 \\ 0 & \alpha_{11} & 0 & 0 & 0 & 0 & 0 & 0 \\ 0 & 0 & \alpha_{22} & 0 & 0 & 0 & 0 & 0 \\ 0 & 0 & 0 & 0 & \alpha_{34} & 0 & 0 & 0 \\ 0 & 0 & 0 & 0 & 0 & \alpha_{45} & 0 & 0 \end{bmatrix}$$

$$\Rightarrow \mathbf{T}_1^l(k) = \mathcal{T}_1(k)\mathbf{T}_1(k),$$

where $\mathbf{T}_1(k)$ is a *scaled orthonormal* matrix and $\mathcal{T}_1(k)$ is the *scaled orthonormalizer* matrix. A scaled orthonormal matrix is defined here as a matrix which is full row rank *and* if its rows are considered as vectors, each row is a scaled unit vector as

$$\mathbf{T}_1(k) = \mathbf{A}(k)\mathbf{U}(k), \tag{4.26}$$

where $\mathbf{A}(k)$ is square, diagonal and nonsingular, while $\mathbf{U}(k)$ is row orthonormal. This is an important property which is essential for the internodal communication theory to be developed later.

The local models, filtering and control are defined using $\mathbf{T}_1^l(k)$, while $\mathbf{T}_1(k)$ is used for internodal communication. Any nodal transformation matrix, full rank or not, can be expressed as a product of two transformation matrices, one of which is a scaled orthonormal matrix. The following example illustrates the case of a nodal transformation matrix which is not full rank, i.e., a transformation that creates a local state vector with redundant states from a global state vector without such states.

Let the local state vector be

$$\mathbf{x}_2(k) = \begin{bmatrix} \alpha_{00}x_0(k) + \alpha_{01}x_1(k) \\\\ \alpha_{10}x_0(k) \\\\ \alpha_{21}x_1(k) \\\\ \alpha_{34}x_4(k) + \alpha_{35}x_5(k) \\\\ \alpha_{45}x_5(k) \end{bmatrix}, \tag{4.27}$$

where the global state vector, $\mathbf{x}(k)$, is as defined before. Thus the general nodal transformation matrix required for this local state vector with redundant states is given by

$$\mathbf{T}_2^l(k) = \begin{bmatrix} \alpha_{00} & \alpha_{01} & 0 & 0 & 0 & 0 & 0 & 0 \\ \alpha_{10} & 0 & 0 & 0 & 0 & 0 & 0 & 0 \\ 0 & \alpha_{21} & 0 & 0 & 0 & 0 & 0 & 0 \\ 0 & 0 & 0 & 0 & \alpha_{34} & \alpha_{35} & 0 & 0 \\ 0 & 0 & 0 & 0 & 0 & \alpha_{45} & 0 & 0 \end{bmatrix}.$$

This matrix, $\mathbf{T}_2^l(k)$, can be expressed in terms of a *scaled* orthonormal matrix $\mathbf{T}_2(k)$ as follows:

$$\mathbf{T}_2^l(k) = \begin{bmatrix} \alpha_{00} & \alpha_{01} & 0 & 0 & 0 & 0 & 0 & 0 \\ \alpha_{10} & 0 & 0 & 0 & 0 & 0 & 0 & 0 \\ 0 & \alpha_{21} & 0 & 0 & 0 & 0 & 0 & 0 \\ 0 & 0 & 0 & 0 & \alpha_{34} & \alpha_{35} & 0 & 0 \\ 0 & 0 & 0 & 0 & 0 & \alpha_{45} & 0 & 0 \end{bmatrix}$$

$$
= \begin{bmatrix} 1 & \frac{\alpha_{01}}{\alpha_{11}} & 0 & 0 \\ \frac{\alpha_{10}}{\alpha_{00}} & 0 & 0 & 0 \\ 0 & \frac{\alpha_{21}}{\alpha_{11}} & 0 & 0 \\ 0 & 0 & 1 & \frac{\alpha_{35}}{\alpha_{45}} \\ 0 & 0 & 0 & 1 \end{bmatrix} \begin{bmatrix} \alpha_{00} & 0 & 0 & 0 & 0 & 0 & 0 & 0 \\ 0 & \alpha_{11} & 0 & 0 & 0 & 0 & 0 & 0 \\ 0 & 0 & 0 & 0 & \alpha_{34} & 0 & 0 & 0 \\ 0 & 0 & 0 & 0 & 0 & \alpha_{45} & 0 & 0 \end{bmatrix}
$$

$$
\Rightarrow \mathbf{T}_2^l(k) = \mathcal{T}_2(k)\mathbf{T}_2(k).
$$

$\mathbf{T}_2(k)$ is thus a scaled orthonormal matrix. Consequently, it can be used for internodal communication in the algorithms derived and discussed in Section 4.4. The difference, however, is that while in the previous example $\mathcal{T}_1(k)$, the scaled orthonormalizer, was full row rank, $\mathcal{T}_2(k)$ is not.

## 4.2.4 Generalizing the Concept

The preceding example can be generalized by considering a general nodal transformation matrix, $\mathbf{T}_g^l(k)$, of size $m \times n$ operating on a general $n$ sized model where $m \le n$. For effective model reduction $m < n$, so that at least one of the columns in $\mathbf{T}_g^l(k)$ must consist of zeros or can be reduced by column operations to satisfy this property.

$$
\mathbf{T}_g^l(k) = \begin{bmatrix} \alpha_{00} & \cdot\cdot & \alpha_{0j} & \cdot\cdot & \alpha_{0(n-2)} & 0 \\ & & \cdot & & \cdot & \cdot \\ \alpha_{i0} & \cdot\cdot & \alpha_{ij} & \cdot\cdot & \alpha_{i(n-2)} & 0 \\ & & \cdot & & \cdot & \cdot \\ \alpha_{(m-1)0} & \cdot\cdot & \alpha_{(m-1)j} & \cdot\cdot & \alpha_{(m-1)(n-2)} & 0 \end{bmatrix}
$$

$$
= \begin{bmatrix} 1 & \cdot\cdot & \frac{\alpha_{0j}}{\alpha_{ij}} & \cdot\cdot & \frac{\alpha_{0(n-2)}}{\alpha_{(m-1)(n-2)}} \\ & & \cdot & & \cdot \\ \frac{\alpha_{i0}}{\alpha_{00}} & \cdot\cdot & 1 & \cdot\cdot & \frac{\alpha_{i(n-2)}}{\alpha_{(m-1)(n-2)}} \\ & & \cdot & & \cdot \\ \frac{\alpha_{(m-1)0}}{\alpha_{ij}} & \cdot\cdot & \frac{\alpha_{(m-1)j}}{\alpha_{ij}} & \cdot\cdot & 1 \end{bmatrix} \times
$$

$$\begin{bmatrix} \alpha_{00} & . . & 0 & . . & 0 & 0 \\ & & & & . & . \\ . & & . & & . & . \\ . & & . & & . & . \\ 0 & . . & \alpha_{ij} & . . & 0 & 0 \\ . & & . & & . & . \\ . & & . & & . & . \\ 0 & . . & 0 & . . & \alpha_{(m-1)(n-2)} & 0 \end{bmatrix}$$

$$\Rightarrow \mathbf{T}_g^l(k) = \mathcal{T}_g(k)\mathbf{T}_g(k),$$

where $\mathbf{T}_g(k)$ is always a scaled row orthonormal matrix, while the scaled orthonormalizer, $\mathcal{T}_g(k)$, is not necessarily full row rank. This effectively generalizes the result to any model reductive nodal transformation applied to a general system.

### 4.2.5   Choice of Transformation Matrices

The internodal transformation theory developed in this book is generalized such that it applies to any nodal transformation matrices $\mathbf{T}_i(k)$ and $\mathbf{T}_j(k)$. However, in any practical system these nodal transformation matrices should be chosen so that they correspond to non-zero local state vectors $\mathbf{x}_i(k)$ and $\mathbf{x}_j(k)$, respectively. The local state vectors should contain all states relevant to the node. Relevance of states is defined by the global state transition matrix $\mathbf{F}(k)$, which expresses the way states affect each other from one time step to the next. If $\mathbf{F}(k)$ is known or easily derived, then $\mathbf{T}_i(k)$ and $\mathbf{T}_j(k)$ are obtained directly from it. Alternatively, the global system can be run at least once to obtain $\mathbf{F}(k)$, which is then used to compute the nodal transformation matrices. These matrices are not necessarily scaled orthonormal.

### 4.2.6   Distribution of Models

At each node, reduced order models which include the local state transition matrix and the process noise model are used. These can be calculated locally at each node as functions of the nodal transformation matrix. They are derived in such a way as to ensure that local estimates and control signals are exactly the same as in the global system [20], [113]. Decentralization of the control signal and its associated models is described in Chapter 5.

The new local state transition equation is given by

$$\mathbf{x}_i(k) = \mathbf{F}_i(k)\mathbf{x}_i(k-1) + \mathbf{B}_i(k)\mathbf{u}_i(k-1) + \mathbf{D}_i(k)\mathbf{w}_i(k-1)$$
$$\Leftrightarrow \mathbf{T}_i(k)\mathbf{x}(k) = \mathbf{F}_i(k)\mathbf{T}_i(k-1)\mathbf{x}(k-1) + \mathbf{B}_i(k)\mathbf{u}_i(k-1)$$
$$+ \mathbf{D}_i(k)\mathbf{T}_i(k-1)\mathbf{w}(k-1). \tag{4.28}$$

Pre-multiplying the global state transition Equation 2.1 by $\mathbf{T}_i(k)$ throughout gives

$$\mathbf{T}_i(k)\mathbf{x}(k) = \mathbf{T}_i(k)\mathbf{F}(k)\mathbf{x}(k-1)+\mathbf{T}_i(k)\mathbf{B}(k)\mathbf{u}(k-1)+\mathbf{T}_i(k)\mathbf{D}(k)\mathbf{w}(k-1).$$
$$(4.29)$$

Comparing Equations 4.28 and 4.29 and hence equating the coefficients of $\mathbf{x}(k-1)$ and $\mathbf{w}(k-1)$ gives the following equations:

$$\mathbf{F}_i(k)\mathbf{T}_i(k-1) = \mathbf{T}_i(k)\mathbf{F}(k) \qquad (4.30)$$

$$\mathbf{D}_i(k)\mathbf{T}_i(k-1) = \mathbf{T}_i(k)\mathbf{D}(k). \qquad (4.31)$$

Equations 4.30 and 4.31 are true for any nodal transformation. They are a discrete time version of Sandell's continuous time, necessary and sufficient dynamic *equivalence* condition [112], [113]. In particular, Equation 4.30 states the requirement that $\mathbf{x}_i(k)$ should be a linear combination of certain modes of $\mathbf{x}(k)$, so that the eigenvalues of $\mathbf{F}_i(k)$ are eigenvalues of $\mathbf{F}(k)$ corresponding to the modes of $\mathbf{x}(k)$ retained in $\mathbf{x}_i(k)$.

For a nodal transformation matrix, $\mathbf{T}_i(k-1) \in \mathcal{R}^{m \times n}$, which produces model size reduction without redundant states in the local state vector, it follows that $m \le n$ and rank $\mathbf{T}_i(k-1) = m$. This means that $\mathbf{T}_i(k-1)$ is full row rank and hence it has a *right* inverse which allows the extraction of expressions for the nodal models from the discrete time Sandell's condition.

$$\mathbf{F}_i(k) = \mathbf{T}_i(k)\mathbf{F}(k)\mathbf{T}_i^+(k-1) \qquad (4.32)$$

$$\mathbf{D}_i(k) = \mathbf{T}_i(k)\mathbf{D}(k)\mathbf{T}_i^+(k-1), \qquad (4.33)$$

where $\mathbf{T}_i^+(k-1)$ is the Moore-Penrose generalized inverse covered in Section 4.3. In practice, all nodal transformation matrices that systematically choose states or linear functions of states from the global state vector, discussed in Section 4.2, are full row rank. Thus Equations 4.32 and 4.33 for nodal models hold for the theory developed in this book.

It is however possible to find an arbitrary nodal transformation which creates a local state vector with redundant states from a system without such states. In this case $\mathbf{T}_i(k)$ will not be full row rank. Consequently, the nodal model expressions in Equations 4.32 and 4.33 will not hold. However, Sandell's Equations 4.30 and 4.31 will still be valid. It is important to note that, by definition, any state transition matrix is always *nonsingular* and hence invertible. This is true for any nodal state transition matrix, $\mathbf{F}_i(k)$, irrespective of whether $\mathbf{T}_i(k)$ is full row rank or not.

## 4.3   The Moore-Penrose Generalized Inverse: $\mathbf{T}^+$

In order to distribute systems models and derive an algorithm that defines what information is communicated between nodes, the inverse of the nodal transformation, $\mathbf{T}_j(k)$, is required. In general $\mathbf{T}_j(k)$ is singular or rectangular and hence an ordinary inverse does not exist. As such, use is made of the *generalized* inverse denoted by $\mathbf{T}_j^+(k)$ and also known as the pseudo-inverse of $\mathbf{T}_j(k)$ [66], [102].

There are four conditions that define generalized inverses [103], [109]:

1. $\mathbf{TT}^+\mathbf{T} = \mathbf{T}$

2. $\mathbf{T}^+\mathbf{TT}^+ = \mathbf{T}^+$

3. $\mathbf{TT}^+ = [\mathbf{TT}^+]^H$

4. $\mathbf{T}^+\mathbf{T} = [\mathbf{T}^+\mathbf{T}]^H$.

The last two conditions state that $\mathbf{TT}^+$ and $\mathbf{T}^+\mathbf{T}$ are Hermitian, i.e., they are equal to their complex conjugate transposes. For real matrices these Hermitian transposes reduce to ordinary transposes, i.e., $\mathbf{A}^H = \mathbf{A}^T$. The work presented here involves only real matrices but can be extended to complex matrices.

Several general inverses (*pseudo-inverses*) can be defined depending on which of the four conditions are satisfied:

- $\mathbf{T}^{+_1}$ satisfies condition 1.

- $\mathbf{T}^{+_2}$ satisfies conditions 1 and 2.

- $\mathbf{T}^{+_3}$ satisfies conditions 1, 2 and 3.

- $\mathbf{T}^{+_3^*}$ satisfies conditions 1, 2 and 4.

- $\mathbf{T}^+$ satisfies all four conditions.

$\mathbf{T}^+$, the inverse which satisfies all four conditions, is called the Moore-Penrose generalized inverse. It is the most generalized of all inverses and exists for every matrix. The other four inverses are called *less* constrained general inverses and are related to each other and to the Moore-Penrose inverse as follows:

$$\mathbf{T}^{+_1} \sqsupseteq \mathbf{T}^{+_2} \sqsupseteq \mathbf{T}^{+_3}(\mathbf{T}^{+_3^*}) \sqsupseteq \mathbf{T}^+. \tag{4.34}$$

This shows inclusiveness of the four classes of inverses, where $\mathbf{A} \sqsupseteq \mathbf{B}$ means that the conditions that $\mathbf{B}$ satisfies include those satisfied by $\mathbf{A}$

[103], [109]. In Equation 4.34, equality holds throughout if and only if
**T** is nonsingular. The Moore-Penrose inverse $\mathbf{T}^+$ is the most constrained
of the five inverses and it exists for any matrix. For this reason it is the
generalized inverse used in this book.

## 4.3.1  Properties and Theorems of $\mathbf{T}^+$

The Moore-Penrose generalized inverse has several properties discussed
and proven in the literature. The following are a selected few which are
important in the derivations presented in this chapter. They are presented
here without proofs, as these are readily available [28], [103], [109].

1. $\mathbf{T}^+$ is unique.

2. $\mathbf{T}^+ = \mathbf{T}^{-1}$ for nonsingular **T**.

3. $(\mathbf{T}^+)^+ = \mathbf{T}$ and $\mathbf{0}^+ = \mathbf{0}$.

4. $(k\mathbf{T})^+ = (1/k)\mathbf{T}^+$ for $k \neq 0$.

5. Ker $\mathbf{T}^+$= Ker $\mathbf{T}^T$ and Im $\mathbf{T}^+$= Im $\mathbf{T}^T$. Ker **A** and Im **A** denote the
   *null space* (kernel) and *range* (image) of matrix **A**, respectively.

6. The rank of $\mathbf{T}^+$ equals the rank of **T**.

7. $\mathbf{TT}^+$ and $\mathbf{T}^+\mathbf{T}$ are both Hermitian and idempotent, i.e., they are
   orthogonal projectors. (**A** is idempotent $\Leftrightarrow \mathbf{A}^2 = \mathbf{A}$.)

8. The inverse product law: $[\mathbf{TA}]^+ = \mathbf{A}^+\mathbf{T}^+$ holds if any one or more
   of the following, non-exhaustive, conditions are true.
   a) **T** or $\mathbf{A}^T$ is column orthonormal.
   b) **T** and **A** admit *right* and *left* inverses, respectively.
   c) $\mathbf{A} = \mathbf{T}^T$.
   d) **T** has size $m \times n$, **A** has size $n \times k$ and both matrices have rank $n$.
   e) $\mathbf{T}^T\mathbf{TAA}^T = \mathbf{AA}^T\mathbf{T}^T\mathbf{T}$.
   f) $\mathbf{AA}^T\mathbf{T}^T\mathbf{T}$ and $\mathbf{T}^T\mathbf{TAA}^+$ are both Hermitian.

9. $[\mathbf{UTV}^T]^+ = \mathbf{VT}^+\mathbf{U}^T$ for any *column orthonormal* matrices, **U** and
   **V**. (**U** is column orthonormal $\Leftrightarrow \mathbf{U}^T\mathbf{U} = \mathbf{1}$.)

10. $[\mathbf{UTV}]^+ = \mathbf{V}^T\mathbf{T}^+\mathbf{U}^T$ for any *unitary* matrices, **U** and **V**. (**U** is
    *unitary* $\Leftrightarrow \mathbf{U}^{-1} = \mathbf{U}^T$.)

11. $\mathbf{T}^T\mathbf{VT} = \mathbf{T}^+\mathbf{V}^{-1}\mathbf{T}^{+T}$ where the rank of the $m \times n$ matrix, **T**, is $m$
    where $(m \leq n)$ and **V** is any *nonsingular* matrix.

12. $[\mathbf{UTV}]^+ = \mathbf{V}^+\mathbf{T}^{-1}\mathbf{U}^+$ for nonsingular **T**.

13. **Least Squares Property**: $\mathbf{T}^+$ provides the *least squares* solution of a general equation of the form

$$\mathbf{T}_j\mathbf{x} = \mathbf{x}_j. \tag{4.35}$$

The result obtained is the best approximate solution and is given by

$$\mathbf{x}_g = \mathbf{T}_j^+\mathbf{x}_j. \tag{4.36}$$

It is best in the sense that $\|\mathbf{T}_j\mathbf{x} - \mathbf{x}_j\|$ is minimized.

$$\|\mathbf{T}_j\mathbf{x}_g - \mathbf{x}_j\| = min\|\mathbf{T}_j\mathbf{x} - \mathbf{x}_j\| \tag{4.37}$$

in the Euclidean vector norm $\|\cdot\|$ [28], [109]. In reconstructing $\mathbf{x}_g$ from $\mathbf{x}_j$, $\mathbf{T}^+$ does not lose any information. The Moore-Penrose inverse preserves information when used in mathematical operations.

Two theorems required for the derivation of the internodal transformation algorithms presented in Section 4.4 are proven from the above properties. Only real matrices are of interest although the theorems and proofs can be extended to complex matrices. These theorems are presented here because of their novel nature and centrality to understanding the use of the Moore-Penrose inverse in the algorithms herein.

**THEOREM 4.1**

$$\left[\mathbf{T}^T\mathbf{I}\mathbf{T}\right]^+ = \mathbf{T}^+\mathbf{I}^+\mathbf{T}^{+T}, \tag{4.38}$$

*where the matrix* $\mathbf{T} \in \Re^{m \times n}$ *is full rank and if its rows are considered as vectors, they form a set of scaled orthonormal vectors so that*

$$\mathbf{T} = \mathbf{A}\mathbf{U}.$$

*The matrix* $\mathbf{A}$ *is square, diagonal and nonsingular.* $\mathbf{U}$ *is a row orthonormal matrix which means* $(\mathbf{U}^+ = \mathbf{U}^T)$ *and* $(\mathbf{U}\mathbf{U}^T = \mathbf{1})$, *where* $\mathbf{1}$ *is the identity matrix.* $\mathbf{I}$ *is any real* $m \times m$ *matrix with diagonal, hermitian and idempotent products,* $\mathbf{I}\,\mathbf{I}^+$ *and* $\mathbf{I}^+\mathbf{I}$.

**PROOF**

$$LHS = \left[\mathbf{T}^T\mathbf{I}\mathbf{T}\right]^+$$

$$= \left[(\mathbf{A}\mathbf{U})^T\mathbf{I}(\mathbf{A}\mathbf{U})\right]^+$$

$$= \left[\mathbf{U}^T\mathbf{A}^T\mathbf{I}\mathbf{A}\mathbf{U}\right]^+$$

$$= \left[\mathbf{U}^T\mathbf{A}\mathbf{I}\mathbf{A}\mathbf{U}\right]^+ \qquad (\mathbf{A}^T = \mathbf{A})$$

$$= \mathbf{U}^T \left[ \mathbf{AIA} \right]^+ \mathbf{U} \qquad (Property \quad 9)$$
$$= \mathbf{U}^T \mathbf{A}^+ \mathbf{I}^+ \mathbf{A}^+ \mathbf{U}$$

($\mathbf{A}$ *is diagonal and nonsingular.* $\mathbf{I\,I}^+$ *and* $\mathbf{I}^+\mathbf{I}$ *are diagonal.*)

$$= \left[ \mathbf{U}^+ \mathbf{A}^+ \right] \mathbf{I}^+ \left[ \mathbf{U}^+ \mathbf{A}^+ \right]^T \qquad (\mathbf{U}^+ = \mathbf{U}^T \text{ and } \mathbf{A}^T = \mathbf{A})$$
$$= \left[ \mathbf{AU} \right]^+ \mathbf{I}^+ \left[ \mathbf{AU} \right]^{+T} \qquad (Property \quad 8a)$$
$$= \mathbf{T}^+ \mathbf{I}^+ \mathbf{T}^{+T}$$
$$= RHS$$

## THEOREM 4.2

$$\left[ \mathbf{VI}^+ \mathbf{V}^T \right]^+ = \mathbf{V}^{+T} \mathbf{IV}^+, \tag{4.39}$$

*where, considering the rows and columns of* $\mathbf{V}$ *as vectors, each row or column is either a scaled unit vector or a zero vector* $(\mathbf{0})$, *that is,*

$$\mathbf{V} = \mathbf{AU}.$$

*The matrix* $\mathbf{A}$ *is square, diagonal but not necessarily nonsingular. Any row or column of* $\mathbf{U}$ *is either a unit vector or zero vector. This means that the condition that* $\mathbf{U}^+ = \mathbf{U}^T$ *holds, while the requirement that* $\mathbf{U}\mathbf{U}^T = \mathbf{1}$ *is not necessarily true.* $\mathbf{I}^+$ *is a square diagonal matrix, which is not necessarily nonsingular.*

## PROOF

$$LHS = \left[ \mathbf{VI}^+ \mathbf{V}^T \right]^+$$
$$= \left[ (\mathbf{AU}) \mathbf{I}^+ (\mathbf{AU})^T \right]^+$$
$$= \left[ \mathbf{AUI}^+ \mathbf{U}^T \mathbf{A}^T \right]^+$$
$$= \left[ \mathbf{A} \left[ \mathbf{UI}^+ \mathbf{U}^T \right] \mathbf{A} \right]^+ \qquad (\mathbf{A}^T = \mathbf{A})$$
$$= \mathbf{A}^+ \left[ \mathbf{UI}^+ \mathbf{U}^T \right]^+ \mathbf{A}^+ \qquad (\mathbf{A} \text{ and } \mathbf{UI}^+ \mathbf{U}^T \text{ are diagonal.})$$
$$= \mathbf{A}^+ \left[ \mathbf{U}^{+T} \mathbf{IU}^+ \right] \mathbf{A}^+ \qquad (\mathbf{I}^+ \text{ and } \mathbf{UI}^+ \mathbf{U}^T \text{ are diagonal.})$$
$$= \left[ \mathbf{A}^+ \mathbf{U} \right] \mathbf{I} \left[ \mathbf{U}^+ \mathbf{A}^+ \right] \qquad (\mathbf{U}^+ = \mathbf{U}^T)$$
$$= \left[ \mathbf{A}^{+T} \mathbf{U} \right] \mathbf{I} \left[ \mathbf{U}^+ \mathbf{A}^+ \right] \qquad (\mathbf{A}^T = \mathbf{A})$$
$$= \left[ \mathbf{U}^+ \mathbf{A}^+ \right]^T \mathbf{I} \left[ \mathbf{U}^+ \mathbf{A}^+ \right] \qquad (\mathbf{U}^+ = \mathbf{U}^T)$$
$$= \left[ \mathbf{AU} \right]^{+T} \mathbf{I} \left[ \mathbf{AU} \right]^+$$

(*Diagonal* $\mathbf{A}$ *and orthnormal* $\mathbf{U} \Rightarrow$ *Property* 8 *holds.*)

$$= \mathbf{V}^{+T}\mathbf{I}\mathbf{V}^{+}$$
$$= RHS$$

These two theorems are used to simplify expressions for the internodal transformation matrices derived and discussed in Section 4.4.

### 4.3.2 Computation of $\mathbf{T}^{+}$

An expression for the Moore-Penrose inverse can obtained by rank decomposition [28], [66]. Consider a general matrix, $\mathbf{T} \in \mathcal{F}^{m \times n}$ with rank $\mathbf{T} = r$. Let

$$\mathbf{T} = \mathbf{F}\mathbf{R}^{T}, \qquad \mathbf{F} \in \mathcal{F}^{m \times r}, \qquad \mathbf{R}^{T} \in \mathcal{F}^{r \times n}$$

be a rank decomposition of $\mathbf{T}$, i.e., rank $\mathbf{F} = $ rank $\mathbf{F}^{T} = r$. Noting that the $(r \times r)$ matrices $\mathbf{F}^{T}\mathbf{F}$ and $\mathbf{R}^{T}\mathbf{R}$ are of full rank, and therefore invertible, an expression of the Moore-Penrose inverse, an $n \times m$ matrix is obtained.

$$\mathbf{T}^{+} = \mathbf{R}[\mathbf{R}^{T}\mathbf{R}]^{-1}[\mathbf{F}^{T}\mathbf{F}]^{-1}\mathbf{F}^{T}. \tag{4.40}$$

Two special cases of $\mathbf{T}^{+}$ are worth noting: the *right* and *left* inverses. A matrix $\mathbf{T} \in \mathcal{F}^{m \times n}$ is said to be *right* (respectively, *left*) invertible if there exists a matrix $\mathbf{T}_{R}^{-1}$ (respectively, $\mathbf{T}_{L}^{-1}$) such that

$$\mathbf{T}\mathbf{T}_{R}^{-1} = \mathbf{1}_{m} \qquad (respectively, \mathbf{T}_{L}^{-1}\mathbf{T} = \mathbf{1}_{n}). \tag{4.41}$$

### Right Inverse, $\mathbf{T}_{R}^{-1}$
The Moore-Penrose inverse reduces to the right inverse if

$$m \leq n \text{ and rank } \mathbf{T} = m. \qquad (full\ row\ rank)$$

This means that rank $\mathbf{T} = $ rank $\mathbf{R} = $ rank $\mathbf{F} = m$ and $\mathbf{F} \in \mathcal{F}^{m \times m}$. Since $\mathbf{F}$ is square and full rank, a normal inverse $\mathbf{F}^{-1}$ exists, simplifying the expression for Moore-Penrose inverse.

$$\begin{aligned} \mathbf{T}_{R}^{-1} &= \mathbf{R}[\mathbf{R}^{T}\mathbf{R}]^{-1}[\mathbf{F}^{T}\mathbf{F}]^{-1}\mathbf{F}^{T} \\ &= \mathbf{R}[\mathbf{R}^{T}\mathbf{R}]^{-1}\mathbf{F}^{-1}[\mathbf{F}^{T}]^{-1}\mathbf{F}^{T} \\ &= \mathbf{R}[\mathbf{R}^{T}\mathbf{R}]^{-1}\mathbf{F}^{-1} \\ &= \mathbf{R}[\mathbf{F}(\mathbf{R}^{T}\mathbf{R})]^{-1} \\ &= [\mathbf{F}^{-1}\mathbf{T}]^{T}[\mathbf{F}(\mathbf{R}^{T}\mathbf{R})]^{-1} \\ &= \mathbf{T}^{T}[\mathbf{F}\mathbf{R}^{T}\mathbf{R}\mathbf{F}^{T}]^{-1} \\ &= \mathbf{T}^{T}[\mathbf{T}\mathbf{T}^{T}]^{-1}. \end{aligned}$$

### Left Inverse, $\mathbf{T}_{L}^{-1}$
The Moore-Penrose inverse reduces to the left inverse if

$$m \geq n \text{ and rank } \mathbf{T} = n. \qquad \textit{(full column rank)}$$

This means that rank $\mathbf{T} = $ rank $\mathbf{R} = $ rank $\mathbf{F} = n$ and $\mathbf{R} \in \mathcal{R}^{n \times n}$. Since $\mathbf{R}$ is square and full rank, a normal inverse $\mathbf{F}^{-1}$ exists. This simplifies the expression for Moore-Penrose inverse.

$$
\begin{aligned}
\mathbf{T}_L^{-1} &= \mathbf{R}[\mathbf{R}^T\mathbf{R}]^{-1}[\mathbf{F}^T\mathbf{F}]^{-1}\mathbf{F}^T \\
&= (\mathbf{R}^{-1})^T[\mathbf{F}^T\mathbf{F}]^{-1}\mathbf{F}^T \\
&= [\mathbf{F}^T\mathbf{F}\mathbf{R}^T]^{-1}\mathbf{F}^T \\
&= [\mathbf{F}^T\mathbf{T}]^{-1}\mathbf{R}^{-1}\mathbf{T}^T \\
&= [\mathbf{R}\mathbf{F}^T\mathbf{T}]^{-1}\mathbf{T}^T \\
&= [\mathbf{T}^T\mathbf{T}]^{-1}\mathbf{T}^T.
\end{aligned}
$$

## 4.4 Generalized Internodal Transformation

In distributed decentralized systems, as in fully connected decentralized systems, information must be communicated between nodes if the systems are to give the same estimates as their centralized equivalents. However, in distributed decentralized systems, since they are non-fully connected, a further issue arises: which nodes *need* to communicate and what information *needs* to be sent between them? A resolution of this question leads to minimization of communication both in terms of the number of communication links and the size of messages. This is accomplished in this book by deriving *internodal transformation* theory for both state and information spaces. An internodal transformation matrix optimally maps information (or states) from one information (or state) subspace to another subspace, such that an accurate local interpretation of the information (or states) is effected in the new subspace.

### 4.4.1 State Space Internodal Transformation: $\mathbf{V}_{ji}(k)$

Consider the problem of carrying out minimum variance estimation at node $i$ based on node $j$'s observations. First, the concept of constructing error covariances and state estimates based only on current observations is introduced. The information contribution at node $i$ due to current observations from node $j$ is defined as $\mathbf{i}_i(\mathbf{z}_j(k))$, where the global information contribution due to global observation is expressed as $\mathbf{i}(\mathbf{z}(k)) = \mathbf{i}(k)$. The associated local information matrix is then defined by $\mathbf{I}_i(\mathbf{z}_j(k))$, where the global associated information matrix is given by $\mathbf{I}(\mathbf{z}(k)) = \mathbf{I}(k)$.

The local error covariance at node $i$ based only on current observations from node $j$ is then defined by

$$\mathbf{P}_i(k \mid \mathbf{z}_j(k)) \overset{\triangle}{=} \mathbf{I}_i^+(\mathbf{z}_j(k)). \tag{4.42}$$

It is important to note that when $\mathbf{P}_i(k \mid \mathbf{z}_j(k))$ and $\mathbf{I}_i(\mathbf{z}_j(k))$ are not invertible (*normal inversion*), $\mathbf{P}_i(k \mid \mathbf{z}_j(k))$ is not strictly a covariance matrix and $\mathbf{I}_i(\mathbf{z}_j(k))$ is not strictly a Fisher information matrix. Both are used here for notational convenience.

A local state estimate at node $i$ based only on observations from node $j$ may then be computed from

$$\hat{\mathbf{x}}_i(k \mid \mathbf{z}_j(k)) \overset{\triangle}{=} \mathbf{P}_i(k \mid \mathbf{z}_j(k))\mathbf{i}_i(\mathbf{z}_j(k)). \tag{4.43}$$

The local error covariance and state estimate at node $j$, based only on current observations $\mathbf{z}_j(k)$, are calculated locally without any need for communication.

$$\mathbf{P}_j(k \mid \mathbf{z}_j(k))$$
$$= \mathrm{E}\left[\mathbf{T}_j(k)\left[\mathbf{x}(k) - \hat{\mathbf{x}}(k \mid \mathbf{z}_j(k))\right]\left\{\mathbf{T}_j(k)\left[\mathbf{x}(k) - \hat{\mathbf{x}}(k \mid \mathbf{z}_j(k))\right]\right\}^T \mid \mathbf{z}_j(k)\right]$$
$$= \mathbf{T}_j(k)\mathrm{E}\left[\left\{\mathbf{x}(k) - \hat{\mathbf{x}}(k \mid \mathbf{z}_j(k))\right\}\left\{\mathbf{x}(k) - \hat{\mathbf{x}}(k \mid \mathbf{z}_j(k))\right\}^T \mid \mathbf{z}_j(k)\right]\mathbf{T}_j^T(k)$$
$$= \mathbf{T}_j(k)\mathbf{P}(k \mid \mathbf{z}_j(k))\mathbf{T}_j^T(k)$$
$$= \mathbf{T}_j(k)\mathbf{I}^+(\mathbf{z}_j(k))\mathbf{T}_j^T(k)$$
$$= \mathbf{T}_j(k)\left[\mathbf{H}_j^T(k)\mathbf{R}_j^+(k)\mathbf{H}_j(k)\right]^+\mathbf{T}_j^T(k)$$
$$= \mathbf{T}_j(k)\left[\left\{\mathbf{C}_j(k)\mathbf{T}_j(k)\right\}^T \mathbf{R}_j^+(k)\left\{\mathbf{C}_j(k)\mathbf{T}_j(k)\right\}\right]^+\mathbf{T}_j^T(k). \tag{4.44}$$

This is the general expression for the local error covariance at node $j$, based only on observations, $\mathbf{z}_j(k)$. It is valid for any nodal transformation matrix $\mathbf{T}_j(k)$ and can be simplified for special cases as discussed later.

The local state estimate may then be computed as

$$\hat{\mathbf{x}}_j(k \mid \mathbf{z}_j(k)) = \mathbf{P}_j(k \mid \mathbf{z}_j(k))\mathbf{i}_j(\mathbf{z}_j(k)), \tag{4.45}$$

where the local information contribution is given by

$$\mathbf{i}_j(\mathbf{z}_j(k)) = \left[\mathbf{C}_j^T(k)\mathbf{R}_j^+(k)\right]\mathbf{z}_j(k). \tag{4.46}$$

The *state space* transformation problem is to find a method of mapping the state estimate and covariance of node $j$ to those of node $i$.

$$\hat{\mathbf{x}}_j(k \mid \mathbf{z}_j(k)) \longmapsto \hat{\mathbf{x}}_i(k \mid \mathbf{z}_j(k))$$
$$\mathbf{P}_j(k \mid \mathbf{z}_j(k)) \longmapsto \mathbf{P}_i(k \mid \mathbf{z}_j(k)).$$

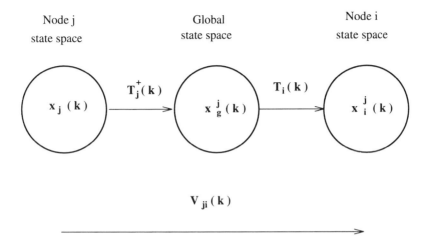

**FIGURE 4.2**
**State Space Transformation**

To solve this problem, the properties of the nodal transformation matrices and their generalized inverses are utilized. $\mathbf{T}_j^+(k)$ takes local states from the state subspace of node $j$, $\mathbf{x}_j(k)$, and expresses them in the global state space, $\mathbf{x}_g^j(k)$. This is achieved because of the key property of the Moore-Penrose inverse, preservation of information. $\mathbf{T}_i(k)$ then picks common states between nodes $j$ and $i$ from this global state space and expresses them in the state subspace of node $i$, $\mathbf{x}_i^j(k)$. This is illustrated in Figure 4.2 and expressed as follows:

$$\mathbf{x}_i^j(k) = \mathbf{T}_i(k)\mathbf{x}_g^j(k)$$
$$= \mathbf{T}_i(k)\mathbf{T}_j^+(k)\mathbf{x}_j(k) \tag{4.47}$$
$$= \mathbf{V}_{ji}(k)\mathbf{x}_j(k), \tag{4.48}$$

where $\mathbf{V}_{ji}(k)$ is defined as the state space internodal transformation matrix,

$$\mathbf{V}_{ji}(k) = \mathbf{T}_i(k)\mathbf{T}_j^+(k). \tag{4.49}$$

The matrix $\mathbf{V}_{ji}(k)$ satisfies the *sufficient* conditions required to transform node $j$'s state estimate $\hat{\mathbf{x}}_j(k \mid \mathbf{z}_j(k))$ to an estimate vector $\hat{\mathbf{x}}_i(k \mid \mathbf{z}_j(k))$ at node $i$. As in the transformation of the actual states, $\mathbf{T}_j^+(k)$ takes the local estimate vector from the $j^{th}$ state subspace, $\hat{\mathbf{x}}_j(k \mid \mathbf{z}_j(k))$, and expresses

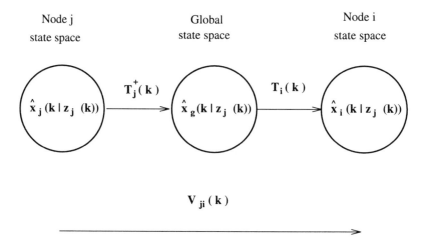

**FIGURE 4.3**
**State Space Transformation (Sufficient Estimation Conditions)**

it in the global state space $\hat{\mathbf{x}}_g(k \mid \mathbf{z}_j(k))$. $\mathbf{T}_i(k)$ then expresses this global estimate in the $i^{th}$ subspace, $\hat{\mathbf{x}}_i(k \mid \mathbf{z}_j(k))$. Figure 4.3 shows this estimate transformation process and can be summarized as

$$\hat{\mathbf{x}}_i(k \mid \mathbf{z}_j(k)) = \mathbf{T}_i(k)\hat{\mathbf{x}}_g(k \mid \mathbf{z}_j(k))$$
$$= \mathbf{T}_i(k)\mathbf{T}_j^+(k)\hat{\mathbf{x}}_j(k \mid \mathbf{z}_j(k)) \qquad (4.50)$$
$$= \mathbf{V}_{ji}(k)\hat{\mathbf{x}}_j(k \mid \mathbf{z}_j(k)). \qquad (4.51)$$

This establishes the transformation of state estimates. The fact that this state space transformation is sufficient, but not necessary, is used in the information space transformation to minimize communication (Section 4.4.2).

The next subproblem is the transformation of state error covariances. From the general definition of covariance (Equation 2.3), the local error covariance at node $i$ with respect to observations from node $j$, $\mathbf{z}_j(k)$, is given by

$$\mathbf{P}_i(k \mid \mathbf{z}_j(k))$$
$$= \mathrm{E}\left[\mathbf{T}_i(k)\left[\mathbf{x}(k) - \hat{\mathbf{x}}(k \mid \mathbf{z}_j(k))\right]\left\{\mathbf{T}_i(k)\left[\mathbf{x}(k) - \hat{\mathbf{x}}(k \mid \mathbf{z}_j(k))\right]\right\}^T \mid \mathbf{z}_j(k)\right]$$
$$= \mathbf{T}_i(k)\mathrm{E}\left[\left\{\mathbf{x}(k) - \hat{\mathbf{x}}(k \mid \mathbf{z}_j(k))\right\}\left\{\mathbf{x}(k) - \hat{\mathbf{x}}(k \mid \mathbf{z}_j(k))\right\}^T \mid \mathbf{z}_j(k)\right]\mathbf{T}_i^T(k)$$
$$\Leftrightarrow \mathbf{P}_i(k \mid \mathbf{z}_j(k)) = \mathbf{T}_i(k)\mathbf{P}(k \mid \mathbf{z}_j(k))\mathbf{T}_i^T(k), \qquad (4.52)$$

where $\mathbf{P}(k \mid \mathbf{z}_j(k))$ is the global error covariance based on observations from node $j$, $\mathbf{z}_j(k)$. The expression for $\mathbf{P}(k \mid \mathbf{z}_j(k))$ is obtained from Equation

4.42 as follows:

$$\mathbf{P}(k \mid \mathbf{z}_j(k)) \triangleq \mathbf{I}^+(\mathbf{z}_j(k))$$
$$= \left[ \mathbf{H}_j^T(k) \mathbf{R}_j^+(k) \mathbf{H}_j(k) \right]^+$$
$$= \left[ \{ \mathbf{C}_j(k) \mathbf{T}_j(k) \}^T \mathbf{R}_j^+(k) \{ \mathbf{C}_j(k) \mathbf{T}_j(k) \} \right]^+$$
$$= \left[ \mathbf{T}_j^T(k) [\mathbf{C}_j^T(k) \mathbf{R}_j^+(k) \mathbf{C}_j(k)] \mathbf{T}_j(k) \right]^+ . \quad (4.53)$$

Substituting Equation 4.53 into Equation 4.52 gives an expression for the local error covariance at node $i$ with respect to observations from node $j$,

$$\mathbf{P}_i(k \mid \mathbf{z}_j(k)) = \mathbf{T}_i(k) \left[ \mathbf{T}_j^T(k) [\mathbf{C}_j^T(k) \mathbf{R}_j^+(k) \mathbf{C}_j(k)] \mathbf{T}_j(k) \right]^+ \mathbf{T}_i^T(k). \quad (4.54)$$

This expression describes a general transformed covariance. It is valid for any nodal transformations $\mathbf{T}_j(k)$ and $\mathbf{T}_i(k)$. If these nodal transformation matrices are scaled orthonormal matrices, Equation 4.54 can be simplified by employing Theorem 4.1.

The expression for $\mathbf{P}_j(k \mid \mathbf{z}_j(k))$ is simpler for scaled orthonormal matrices.

$$\mathbf{P}_j(k \mid \mathbf{z}_j(k)) = \mathbf{T}_j(k) \left[ \{ \mathbf{C}_j(k) \mathbf{T}_j(k) \}^T \mathbf{R}_j^+(k) \{ \mathbf{C}_j(k) \mathbf{T}_j(k) \} \right]^+ \mathbf{T}_j^T(k)$$
$$= \mathbf{T}_j(k) [\mathbf{C}_j(k) \mathbf{T}_j(k)]^+ \mathbf{R}_j(k) [\mathbf{C}_j(k) \mathbf{T}_j(k)]^{+T} \mathbf{T}_j^T(k)$$
$$= \left[ \mathbf{T}_j(k) \mathbf{T}_j^+(k) \right] \mathbf{C}_j^+(k) \mathbf{R}_j(k) \mathbf{C}_j^{+T}(k) \left[ \mathbf{T}_j(k) \mathbf{T}_j^+(k) \right]^T$$
$$= \left[ \mathbf{T}_j(k) \mathbf{T}_j^+(k) \right] \mathbf{C}_j^+(k) \mathbf{R}_j(k) \mathbf{C}_j^{+T}(k) \left[ \mathbf{T}_j(k) \mathbf{T}_j^+(k) \right]^T$$
$$= \mathbf{C}_j^+(k) \mathbf{R}_j(k) \mathbf{C}_j^{+T}(k)$$
$$= \left[ \mathbf{C}_j^T(k) \mathbf{R}_j^+(k) \mathbf{C}_j(k) \right]^+ \quad (4.55)$$
$$= \mathbf{I}_j^+(\mathbf{z}_j(k)). \quad (4.56)$$

Substituting Equation 4.55 in Equation 4.54 gives

$$\mathbf{P}_i(k \mid \mathbf{z}_j(k)) = \mathbf{T}_i(k) \left[ \mathbf{T}_j^T(k) \mathbf{P}_j^+(k \mid \mathbf{z}_j(k)) \mathbf{T}_j(k) \right]^+ \mathbf{T}_i^T(k). \quad (4.57)$$

This illustrates direct transformation of the error covariance from one state space to the other.

Using Theorem 4.1 in Equation 4.57,

$$\mathbf{P}_i(k \mid \mathbf{z}_j(k)) = \mathbf{T}_i(k) \left[ \mathbf{T}_j^+(k) \mathbf{P}_j(k \mid \mathbf{z}_j(k)) \mathbf{T}_j^{+T}(k) \right] \mathbf{T}_i^T(k)$$
$$= \left[ \mathbf{T}_i(k) \mathbf{T}_j^+(k) \right] \mathbf{P}_j(k \mid \mathbf{z}_j(k)) \left[ \mathbf{T}_i(k) \mathbf{T}_j^+(k) \right]^T$$
$$= \mathbf{V}_{ji}(k) \mathbf{P}_j(k \mid \mathbf{z}_j(k)) \mathbf{V}_{ji}^T(k), \quad (4.58)$$

where $\mathbf{V}_{ji}(k)$ is the state space transformation matrix given in Equation 4.49. Comparing the transformed error covariance given in Equation 4.58 with the transformed state estimate in Equation 4.51, it is clear that the two expressions are consistent with the general definition of covariance, that is,

$$\mathbf{P}_i(k \mid \mathbf{z}_j(k)) =$$
$$\mathrm{E}\left[\mathbf{V}_{ji}(k)\left[\mathbf{x}_j(k) - \hat{\mathbf{x}}_j(k \mid \mathbf{z}_j(k))\right]\left\{\mathbf{V}_{ji}(k)\left[\mathbf{x}_j(k) - \hat{\mathbf{x}}_j(k \mid \mathbf{z}_j(k))\right]\right\}^T \mid \mathbf{z}_j(k)\right].$$

This consistency should always hold since by definition the state estimate and the covariance are measures of the the same quantity, the true state. It is useful to note that the derivation of $\mathbf{V}_{ji}(k)$ places no constraints on $\mathbf{C}_j(k)$ and $\mathbf{R}_j(k)$.

### 4.4.2   Information Space Internodal Transformation: $\mathbf{T}_{ji}(k)$

Given the advantages of information space over state space discussed in Chapter 2, it is useful to develop an internodal transformation technique which allows the transformation of information from the $j^{th}$ information subspace to that of node $i$. Instead of transforming and communicating state estimates and error covariances, only information derived from observations is transformed and communicated. A further advantage of this approach, over state space transformation, is that it satisfies minimum (necessary) transformation conditions. Hence, the resulting network topology has minimized internodal communication. The objective is to find a way of directly transforming the information contribution and its associated matrix from node $j$ to the corresponding information contribution and associated matrix at node $i$, given only node $j$ observations $\mathbf{z}_j(k)$.

$$\mathbf{i}_j(\mathbf{z}_j(k)) \longmapsto \mathbf{i}_i(\mathbf{z}_j(k))$$
$$\mathbf{I}_j(\mathbf{z}_j(k)) \longmapsto \mathbf{I}_i(\mathbf{z}_j(k)).$$

The transformation of the associated information matrix $\mathbf{I}_j(\mathbf{z}_j(k))$ into node $i$ information space follows from Equation 4.54.

$$\mathbf{I}_i(\mathbf{z}_j(k)) = \left[\mathbf{T}_i(k)\left[\mathbf{T}_j^T(k)\mathbf{I}_j(\mathbf{z}_j(k))\mathbf{T}_j(k)\right]^+ \mathbf{T}_i^T(k)\right]^+. \qquad (4.59)$$

This transformation holds for any nodal transformations $\mathbf{T}_i(k)$, $\mathbf{T}_j(k)$ and associated information matrix $\mathbf{I}_j(\mathbf{z}_j(k))$.

Next, the transformation of the information contribution, $\mathbf{i}_j(\mathbf{z}_j(k))$ to $\mathbf{i}_i(\mathbf{z}_j(k))$ is considered. The required process of transformation is illustrated in Figure 4.4. $\mathbf{I}_j^+(\mathbf{z}_j(k))$ transforms nodal information from the $j^{th}$ information subspace into state estimates in the $j^{th}$ state subspace. $\mathbf{V}_{ji}(k)$

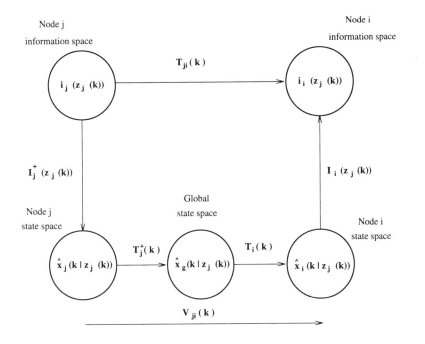

**FIGURE 4.4**
**Information Space Transformation**

picks the common state estimates between these created estimates and the $i^{th}$ state subspace and transforms them from the $j^{th}$ state subspace into the $i^{th}$ state subspace. Finally, $\mathbf{I}_i(\mathbf{z}_j(k))$ changes these state estimates from the $i^{th}$ state subspace into the $i^{th}$ information space.

The derivation of the transformation matrix is shown in Figure 4.4 and proceeds as follows:

$$
\begin{aligned}
\mathbf{i}_i(\mathbf{z}_j(k)) &= \mathbf{I}_i(\mathbf{z}_j(k))\hat{\mathbf{x}}_i(k \mid \mathbf{z}_j(k)) \\
&= \mathbf{I}_i(\mathbf{z}_j(k))\mathbf{V}_{ji}(k)\hat{\mathbf{x}}_j(k \mid \mathbf{z}_j(k)) \\
&= \mathbf{I}_i(\mathbf{z}_j(k))\mathbf{V}_{ji}(k)\mathbf{I}_j^+(\mathbf{z}_j(k))\mathbf{i}_j(\mathbf{z}_j(k)) \\
&= \mathbf{T}_{ji}(k)\mathbf{i}_j(\mathbf{z}_j(k)).
\end{aligned}
\tag{4.60}
$$

$\mathbf{T}_{ji}(k)$ is defined as the *information space* internodal transformation matrix. It picks and maps *relevant* information contributions from the $j^{th}$ information subspace to the $i^{th}$ information subspace.

$$
\begin{aligned}
\mathbf{T}_{ji}(k) &= \mathbf{I}_i(\mathbf{z}_j(k))\left[\mathbf{T}_i(k)\mathbf{T}_j^+(k)\right]\mathbf{I}_j^+(\mathbf{z}_j(k)) \\
&= \mathbf{I}_i(\mathbf{z}_j(k))\mathbf{V}_{ji}(k)\mathbf{I}_j^+(\mathbf{z}_j(k)).
\end{aligned}
\tag{4.61}
$$

The matrix $\mathbf{V}_{ji}(k)$ represents the state space transformation,

$$\mathbf{V}_{ji}(k) = \mathbf{T}_i(k)\mathbf{T}_j^+(k),$$

and $\mathbf{I}_i(\mathbf{z}_j(k))$ is the transformed associated information matrix,

$$\mathbf{I}_i(\mathbf{z}_j(k)) = \left[\mathbf{T}_i(k)\left[\mathbf{T}_j^T(k)\mathbf{I}_j(\mathbf{z}_j(k))\mathbf{T}_j(k)\right]^+\mathbf{T}_i^T(k)\right]^+. \qquad (4.62)$$

Equation 4.61 is the most generalized form of the information space internodal transformation matrix, $\mathbf{T}_{ji}(k)$. This result holds for any nodal transformation paradigm (model defined or arbitrary), any nodal observation matrices and any noise models, where the choice of the nodal transformation matrices $\mathbf{T}_i(k)$ and $\mathbf{T}_j(k)$ satisfy the conditions discussed in Section 4.2.5.

## 4.5    Special Cases of $\mathbf{T}_{ji}(k)$

Several simplified expressions of the information space transformation matrix can be obtained by imposing constraints on the nodal transformation matrices, observation models and noises. Although the generalized result in Equation 4.61 is always true and applicable, it might be unnecessary for some specific cases in which the choice of nodal transformations or observation models are constrained in some way [20], [22], [113]. The following sections discuss such special cases.

### 4.5.1    Scaled Orthonormal $\mathbf{T}_i(k)$ and $\mathbf{T}_j(k)$

The broadest case is when the nodal transformations, $\mathbf{T}_j(k)$ and $\mathbf{T}_i(k)$, are scaled orthonormal without further restrictions on local observation matrices and noise models. This case is of particular interest since the condition satisfied is the same as the one satisfied by the systematically derived transformations discussed in Section 4.2. Application of Theorem 4.1 simplifies the expression for $\mathbf{I}_i(\mathbf{z}_j(k))$.

$$\begin{aligned}
\mathbf{I}_i(\mathbf{z}_j(k)) &= \left[\mathbf{T}_i(k)\left[\mathbf{T}_j^T(k)\mathbf{I}_j(\mathbf{z}_j(k))\mathbf{T}_j(k)\right]^+\mathbf{T}_i^T(k)\right]^+ \\
&= \left[\left\{\mathbf{T}_i(k)\mathbf{T}_j^+(k)\right\}\mathbf{I}_j^+(\mathbf{z}_j(k))\left\{\mathbf{T}_i(k)\mathbf{T}_j^+(k)\right\}^T\right]^+ \\
&= \left[\mathbf{V}_{ji}(k)\mathbf{I}_j^+(\mathbf{z}_j(k))\mathbf{V}_{ji}^T(k)\right]^+. \qquad (4.63)
\end{aligned}$$

Substituting this result in Equation 4.61 gives an expression of the inter-nodal transformation matrix.

$$\mathbf{T}_{ji}(k) = \left[ \mathbf{V}_{ji}(k)\mathbf{I}_j^+(\mathbf{z}_j(k))\mathbf{V}_{ji}^T(k) \right]^+ \mathbf{V}_{ji}(k)\mathbf{I}_j^+(\mathbf{z}_j(k)). \qquad (4.64)$$

This result is valid for all the derived nodal transformations and allows linear combinations of observations. Further subcases of Equation 4.64 can be deduced by imposing more constraints.

## 4.5.2  Diagonal $\mathbf{I}_j^+(\mathbf{z}_j(k))$

If the restriction that $\mathbf{I}_j^+(\mathbf{z}_j(k))$ is diagonal is imposed, so that linear combinations of observations are not possible at nodes, Equation 4.64 can be simplified using Theorem 4.2.

$$\left[ \mathbf{V}_{ji}(k)\mathbf{I}_j^+(\mathbf{z}_j(k))\mathbf{V}_{ji}^T(k) \right]^+ = \mathbf{V}_{ji}^{+T}(k)\mathbf{I}_j(\mathbf{z}_j(k))\mathbf{V}_{ji}^+(k) \qquad (4.65)$$

$$\Rightarrow \mathbf{T}_{ji}(k) = \mathbf{V}_{ji}^{+T}(k)\mathbf{I}_j(\mathbf{z}_j(k))\mathbf{V}_{ji}^+(k)\mathbf{V}_{ji}(k)\mathbf{I}_j^+(\mathbf{z}_j(k))$$

$$= \mathbf{V}_{ji}^{+T}(k)\mathbf{V}_{ji}^+(k)\mathbf{V}_{ji}(k)\mathbf{I}_j(\mathbf{z}_j(k))\mathbf{I}_j^+(\mathbf{z}_j(k))$$

$$= \left[ \mathbf{V}_{ji}^+(k)\mathbf{V}_{ji}(k)\mathbf{V}_{ji}(k) \right]^T \mathbf{I}_j(\mathbf{z}_j(k))\mathbf{I}_j^+(\mathbf{z}_j(k))$$

$$= \mathbf{V}_{ji}^{+T}(k)\mathbf{I}_j(\mathbf{z}_j(k))\mathbf{I}_j^+(\mathbf{z}_j(k)). \qquad (4.66)$$

This result is established by employing the property that the product $\left[\mathbf{V}_{ji}^+(k)\mathbf{V}_{ji}(k)\right]$ is both diagonal and hermitian. Since most applications do not have nodal linear combinations of observations, the above formula is practically useful.

## 4.5.3  Nonsingular and Diagonal $\mathbf{I}_j^+(\mathbf{z}_j(k))$

When $\mathbf{I}_j^+(\mathbf{z}_j(k))$ is nonsingular and diagonal, all local states are observed locally as independent observations. Since $\mathbf{I}_j^+(\mathbf{z}_j(k))$ is nonsingular it has an ordinary inverse, hence Equation 4.66 reduces to

$$\mathbf{T}_{ji}(k) = \mathbf{V}_{ji}^{+T}(k), \qquad (4.67)$$

which means the information internodal transformation depends only on the state space transformation. From Equation 4.63, the expression for the transformed associated information matrix is then simplified.

$$\mathbf{I}_i(\mathbf{z}_j(k)) = \left[ \mathbf{V}_{ji}(k)\mathbf{I}_j^+(\mathbf{z}_j(k))\mathbf{V}_{ji}^T(k) \right]^+$$

$$= \mathbf{V}_{ji}^{+T}(k)\mathbf{I}_j(\mathbf{z}_j(k))\mathbf{V}_{ji}^+(k) \qquad (\textit{Property } 12)$$

$$= \mathbf{T}_{ji}(k)\mathbf{I}_j(\mathbf{z}_j(k))\mathbf{T}_{ji}^T(k), \qquad (4.68)$$

where the transformed information contribution is given by

$$\mathbf{i}_i(\mathbf{z}_j(k)) = \mathbf{T}_{ji}(k)\mathbf{i}_j(\mathbf{z}_j(k)). \qquad (4.69)$$

Equations 4.68 and 4.69 give a computationally convenient form of information transformation. They clearly illustrate the consistency between the transformed information contribution and its associated information matrix. This consistency always holds because $\mathbf{i}_j(\mathbf{z}_j(k))$ and $\mathbf{I}_i(\mathbf{z}_j(k))$ are measures of information due to the same local observation $\mathbf{z}_j(k)$. The constraints imposed to obtain Equation 4.68 are satisfied by some interesting practical problems, so this form of information transformation is also practically useful.

### 4.5.4  Row Orthonormal $\mathbf{C}_j(k)$ and Nonsingular $\mathbf{R}_j(k)$

This is the case where no scaling of observed individual states takes place and the noise covariance $\mathbf{R}_j(k)$ is nonsingular. In this case not all local states are necessarily observed locally and hence $\mathbf{T}_{ji}(k)$ can be simplified further. Consider the expression of $\mathbf{I}_j^+(\mathbf{z}_j(k))$.

$$
\begin{aligned}
\mathbf{I}_j^+(\mathbf{z}_j(k)) &= \left[\mathbf{C}_j^T(k)\mathbf{R}_j^{-1}(k)\mathbf{C}_j(k)\right]^+ \\
&= \mathbf{C}_j^T(k)\mathbf{R}_j(k)\mathbf{C}_j(k) \quad (Property\ \ 9) \\
\Rightarrow \mathbf{I}_j(\mathbf{z}_j(k))\mathbf{I}_j^+(\mathbf{z}_j(k)) &= \left[\mathbf{C}_j^T(k)\mathbf{R}_j^{-1}(k)\mathbf{C}_j(k)\right]\left[\mathbf{C}_j^T(k)\mathbf{R}_j(k)\mathbf{C}_j(k)\right] \\
&= \mathbf{C}_j^T(k)\mathbf{R}_j^{-1}(k)\left[\mathbf{C}_j(k)\mathbf{C}_j^T(k)\right]\mathbf{R}_j(k)\mathbf{C}_j(k) \\
&\qquad (Column\ \ orthonormal) \\
&= \mathbf{C}_j^T(k)\mathbf{C}_j(k). \qquad (4.70)
\end{aligned}
$$

Substituting this result in Equation 4.66 gives

$$
\begin{aligned}
\mathbf{T}_{ji}(k) &= \mathbf{V}_{ji}^{+T}(k)\mathbf{C}_j^T(k)\mathbf{C}_j(k) \\
&= \left[\mathbf{T}_i(k)\mathbf{T}_j^+(k)\right]^{+T}\mathbf{C}_j^T(k)\mathbf{C}_j(k) \\
&= \left[\mathbf{T}_j(k)\mathbf{T}_i^+(k)\right]^T\mathbf{C}_j^T(k)\mathbf{C}_j(k) \quad (Property\ \ 8b) \\
&= \left[\mathbf{C}_j^T(k)\mathbf{C}_j(k)\mathbf{T}_j(k)\mathbf{T}_i^+(k)\right]^T \\
&= \left[\mathbf{C}_j^T(k)\mathbf{H}_j(k)\mathbf{T}_i^+(k)\right]^T \\
&= \left[\mathbf{S}_j(k)\mathbf{T}_i^+(k)\right]^T \\
&= \mathbf{T}_i^{+T}(k)\mathbf{S}_j^T(k), \qquad (4.71)
\end{aligned}
$$

where

$$\mathbf{S}_j(k) = \mathbf{C}_j^T(k)\mathbf{H}_j(k), \qquad (4.72)$$

defines the nodal observation transformation matrix, that is, the matrix which indicates those local states that are observed.

Equation 4.72 gives the *necessary* condition between the local observation models. This result, which is independently derived here, corresponds to the one proposed by Berg [20]. However, as illustrated, it is just a subcase of the entire problem.

## 4.5.5 Row Orthonormal $T_i(k)$ and $T_j(k)$

In this case all rows of a nodal transformation matrix are orthogonal unit vectors. This means that $\mathbf{T}_j(k)\mathbf{T}_j^T(k) = \mathbf{1}$ and $\mathbf{T}_i(k)\mathbf{T}_i^T(k) = \mathbf{1}$. The nodal transformation matrices pick unscaled states from the global state vector and no local state is a linear combination of global states. This has the effect of simplifying the state space transformation matrix to

$$\mathbf{V}_{ji}(k) = \mathbf{T}_i(k)\mathbf{T}_j^T(k). \tag{4.73}$$

This expression can be substituted into the other cases, providing further simplification. For example, if all local states are observed, then substituting Equation 4.73 into Equation 4.67 gives

$$\begin{aligned} \mathbf{T}_{ji}(k) &= \mathbf{V}_{ji}(k) \\ &= \mathbf{T}_i(k)\mathbf{T}_j^T(k). \end{aligned} \tag{4.74}$$

Similarly, if $\mathbf{C}_j(k)$ is row orthonormal and $\mathbf{R}_j(k)$ is nonsingular, this leads to

$$\mathbf{T}_{ji}(k) = \mathbf{T}_i(k)\mathbf{S}_j^T(k). \tag{4.75}$$

## 4.5.6 Reconstruction of Global Variables

In a system with model distribution it might be necessary to reconstruct the global state vector or parts of it in a global state space. This might be for overall system monitoring and information extraction from a node at an arbitrary position in the network. This problem requires the solution to a general equation of the form

$$\mathbf{T}_i(k)\mathbf{x}(k) = \mathbf{x}_i(k). \tag{4.76}$$

The Moore-Penrose inverse has an elegant use in the study of equations of this nature. Application of Property 13 (Section 4.3.1) provides the best approximate solution of this equation, given by

$$\mathbf{x_g}(k) = \mathbf{T}_i^+(k)\mathbf{x}_i(k). \tag{4.77}$$

It is best in the sense that $\|\mathbf{T}_i(k)\mathbf{x}(k) - \mathbf{x}_i(k)\|$ is minimized:

$$\|\mathbf{T}_i(k)\mathbf{x}_g(k) - \mathbf{x}_i(k)\| = min\|\mathbf{T}_i(k)\mathbf{x}(k) - \mathbf{x}_i(k)\| \tag{4.78}$$

in the Euclidean vector norm $\| \cdot \|$. The vector $\mathbf{x_g}(k)$ is also known as a least squares solution of the system described by Equation 4.76.

For the model defined nodal transformation matrices, derived and discussed in Section 4.2, this solution has an important geometrical interpretation. $\mathbf{T}_i^+(k)$ reconstructs all locally relevant states for node $i$ in the global space. Hence $\mathbf{x_g}(k)$ is a global vector consisting of unscaled global states relevant to node $i$ and a *zero* in place of any state irrelevant to node $i$. Similarly, global models are reconstructed giving only the locally relevant components and zeros elsewhere. For this reason, it is clear that $\mathbf{T}_i(k)$ preserves information when applied to transform systems from one subspace to another.

## 4.6   Distributed and Decentralized Filters

Model distribution and the internodal transformation theory which is derived from it can be used to establish scalable, non-fully connected decentralized filters in both state and information spaces. This is achieved by local internodal communication and use of reduced order local models. Such filters do not have the drawbacks of fully connected decentralized estimation algorithms. The network topology of communicating sensor nodes is model defined (dependent on $\mathbf{F}(k)$) and hence there is no need to propagate information between two unconnected nodes. This is because for any two such nodes, information from either node is irrelevant to the other. For this reason, the need for channel filters does not arise. The derivation of distributed and decentralized filters follows from that of their fully connected decentralized equivalents by applying internodal communication theory developed in the previous section.

### 4.6.1   The Distributed and Decentralized Kalman Filter (DDKF)

In state space, the internodal transformation, $\mathbf{V}_{ji}(k)$, defines which sensor nodes need to communicate and explicitly indicates which pieces of information have to be communicated between any two nodes. It can be used to remove the requirement for a fully connected network in the DKF. This produces a scalable decentralized estimation algorithm which is algebraically equivalent to the DKF: the distributed and decentralized Kalman filter (**DDKF**).

#### Prediction

As is the case for the DKF, nodal state and variance prediction equations for the DDKF directly follow from those of the Kalman filter. However, unlike the DKF, the local state transition, noise and control models are different from global ones. This is because of model distribution, which produces reduced order models. Hence, the state and covariance predictions are computed as follows:

$$\hat{\mathbf{x}}_i(k \mid k-1) = \mathbf{F}_i(k)\hat{\mathbf{x}}_i(k-1 \mid k-1) + \mathbf{B}_i(k)\mathbf{u}_i(k-1) \quad (4.79)$$
$$\mathbf{P}_i(k \mid k-1) = \mathbf{F}_i(k)\mathbf{P}_i(k-1 \mid k-1)\mathbf{F}_i^T(k) + \mathbf{Q}_i(k). \quad (4.80)$$

In this way prediction is carried out locally before any communication with the other sensor nodes. The predicted quantities are then used, after internodal communication, to produce estimates.

**Estimation**

The estimation equations are obtained by applying state space internodal transformations. Local estimates, based only on observations, of covariance and state are transformed from one nodal state subspace to the other. They are then assimilated locally to produce state and covariance estimates.

$$\hat{\mathbf{x}}_i(k \mid k) =$$
$$\mathbf{P}_i(k \mid k) \left\{ \mathbf{P}_i^{-1}(k \mid k-1)\hat{\mathbf{x}}_i(k \mid k-1) + \sum_{j=1}^{N} \left[ \mathbf{P}_i^+(k \mid \mathbf{z}_j(k))\hat{\mathbf{x}}_i(k \mid \mathbf{z}_j(k)) \right] \right\}$$

$$\mathbf{P}_i(k \mid k) = \left[ \mathbf{P}_i^{-1}(k \mid k-1) + \sum_{j=1}^{N} \mathbf{P}_i^+(k \mid \mathbf{z}_j(k)) \right]^+,$$

where the transformed covariance and state estimates are given by

$$\mathbf{P}_i(k \mid \mathbf{z}_j(k)) = \mathbf{T}_i(k) \left[ \mathbf{T}_j^T(k)\mathbf{P}_j^+(k \mid \mathbf{z}_j(k))\mathbf{T}_j(k) \right]^+ \mathbf{T}_i^T(k) \quad (4.81)$$
$$\hat{\mathbf{x}}_i(k \mid \mathbf{z}_j(k)) = \mathbf{V}_{ji}(k)\hat{\mathbf{x}}_j(k \mid \mathbf{z}_j(k)). \quad (4.82)$$

The state space internodal transformation matrix is obtained from

$$\mathbf{V}_{ji}(k) = \mathbf{T}_i(k)\mathbf{T}_j^+(k). \quad (4.83)$$

Local covariance and state estimates at node $j$ are computed from local observations,

$$\mathbf{P}_j(k \mid \mathbf{z}_j(k)) = \mathbf{T}_j(k) \left[ \{\mathbf{C}_j(k)\mathbf{T}_j(k)\}^T \mathbf{R}_j^+(k) \{\mathbf{C}_j(k)\mathbf{T}_j(k)\} \right]^+ \mathbf{T}_j^T(k)$$
$$\hat{\mathbf{x}}_j(k \mid \mathbf{z}_j(k)) = \mathbf{P}_j(k \mid \mathbf{z}_j(k))\mathbf{i}_j(\mathbf{z}_j(k)).$$

If the nodal transformation matrices, $\mathbf{T}_i(k)$ and $\mathbf{T}_j(k)$, are scaled orthonormal, then the assimilation equations are simplified by the following substitutions:

$$\mathbf{P}_i(k \mid \mathbf{z}_j(k)) = \mathbf{V}_{ji}(k)\mathbf{P}_j(k \mid \mathbf{z}_j(k))\mathbf{V}_{ji}^T(k) \qquad (4.84)$$

$$\mathbf{P}_j(k \mid \mathbf{z}_j(k)) = \left[\mathbf{C}_j^T(k)\mathbf{R}_j^+(k)\mathbf{C}_j(k)\right]^+ . \qquad (4.85)$$

Comparing Equations 4.85 and 4.82 shows consistency between these two equations and the general definition of covariance as expressed in Equation 2.3. Estimates of individual states obtained from the DDKF algorithm are exactly the same as those obtained from the DKF or from an equivalent centralized Kalman filter algorithm.

However, the advantages of the DDKF over the DKF include fewer communication links, smaller information messages, reduced model sizes and improved system scalability. Such a model defined, non-fully connected, estimation topology does not require propagation of information between unconnected nodes.

## 4.6.2   The Distributed and Decentralized Information Filter (DDIF)

As discussed in Chapter 3, information variables are easy to initialize, distribute and decentralize. It is computationally easier for nodes to exchange information about states rather than communicate actual state estimates. Consequently, information space seems the most natural framework to carry out scalable multisensor estimation. Moreover, the information space internodal transformation matrix $\mathbf{T}_{ji}(k)$ satisfies only the *necessary* observation transformation conditions. Since this matrix defines which sensor nodes need to communicate and explicitly indicates which pieces of information any such nodes have to share, it effectively *minimizes* communication both in terms of message size and number of communication links. In this way, the communication requirements are reduced even further than in the DDKF. The filter based on $\mathbf{T}_{ji}(k)$ is defined as the distributed and decentralized Information filter (**DDIF**). It is a scalable decentralized estimation algorithm in linear information space.

### Prediction

The prediction equations of the DDIF are in the same format as those of the DIF and they are derived in a similar way. The difference is that local system models are not the same as global ones. Each node computes local predictions based on previous local information estimates and *local* system models as follows:

$$\hat{\mathbf{y}}_i(k \mid k - 1) = \mathbf{L}_i(k \mid k - 1)\hat{\mathbf{y}}_i(k - 1 \mid k - 1) \qquad (4.86)$$

$$\mathbf{Y}_i(k \mid k-1) = \left[ \mathbf{F}_i(k) \mathbf{Y}_i^{-1}(k-1 \mid k-1) \mathbf{F}_i^T(k) + \mathbf{Q}_i(k) \right]^{-1}, \quad (4.87)$$

where the local propagation coefficient, independent of the observations made, is given by

$$\mathbf{L}_i(k \mid k-1) = \mathbf{Y}_i(k \mid k-1) \mathbf{F}_i(k) \mathbf{Y}_i^{-1}(k-1 \mid k-1). \quad (4.88)$$

These predictions are of reduced state order, consisting only of locally relevant states at node $i$. The predictions are then used to compute local estimates.

**Estimation**

Not all sensor nodes in the network communicate. The communicating nodes exchange only relevant information which is in the form of transformed information contributions and associated information matrices. At any one node, communicated information, local information and predictions are assimilated to produce local information estimates which are of reduced order.

$$\hat{\mathbf{y}}_i(k \mid k) = \hat{\mathbf{y}}_i(k \mid k-1) + \sum_{j=1}^{N} \left[ \mathbf{T}_{ji}(k) \mathbf{i}_j(\mathbf{z}_j(k)) \right] \quad (4.89)$$

$$\mathbf{Y}_i(k \mid k) = \mathbf{Y}_i(k \mid k-1) + \sum_{j=1}^{N} \mathbf{I}_i(\mathbf{z}_j(k)). \quad (4.90)$$

The transformed associated information matrix and the information space internodal transformation are given by

$$\mathbf{I}_i(\mathbf{z}_j(k)) = \left[ \mathbf{T}_i(k) \left[ \mathbf{T}_j^T(k) \mathbf{I}_j(\mathbf{z}_j(k)) \mathbf{T}_j(k) \right]^+ \mathbf{T}_i^T(k) \right]^+ \quad (4.91)$$

$$\mathbf{T}_{ji}(k) = \mathbf{I}_i(\mathbf{z}_j(k)) \mathbf{V}_{ji}(k) \mathbf{I}_j^+(\mathbf{z}_j(k)). \quad (4.92)$$

For scaled orthonormal transformation matrices, $\mathbf{T}_i(k)$ and $\mathbf{T}_j(k)$, and nonsingular, diagonal information matrix $\mathbf{I}_j^+(\mathbf{z}_j(k))$, the assimilation Equation 4.90 reduces to

$$\mathbf{Y}_i(k \mid k) = \mathbf{Y}_i(k \mid k-1) + \sum_{j=1}^{N} \left[ \mathbf{T}_{ji}(k) \mathbf{I}_j(\mathbf{z}_j(k)) \mathbf{T}_{ji}^T(k) \right], \quad (4.93)$$

where

$$\mathbf{T}_{ji}(k) = \mathbf{V}_{ji}^{+T}(k). \quad (4.94)$$

Comparing Equations 4.93 and 4.89 clearly illustrates the consistency between the information state and matrix estimates. The two estimates are essentially measures of information about the same parameter, state vector $\mathbf{x}_i(k)$. These two equations can be further simplified by employing special cases of $\mathbf{T}_{ji}(k)$ as discussed in Section 4.5.

### 4.6.3   The Distributed and Decentralized Extended Kalman Filter (DDEKF)

A scalable decentralized estimation scheme has been developed for both linear state and information spaces. The next step is to extend these algorithms to deal with nonlinear estimation problems. In state space such an algorithm can be obtained by distributing the system models in the DEKF and employing internodal local communication based on the state space transformation matrix $\mathbf{V}_{ji}(k)$. The result is a state space, non-fully connected network of communicating sensor nodes, each taking nonlinear observations, while its state vector evolves nonlinearly. This is the distributed and decentralized extended Kalman filter (**DDEKF**).

**Prediction**

Other than having local models that are different from global ones, due to model distribution, the DDEKF prediction equations are similar to those of the DEKF. Predictions depend on previous local state estimates and reduced order linearized models.

$$\hat{\mathbf{x}}_i(k \mid k-1) = \mathbf{f}_i\left(k, \hat{\mathbf{x}}_i(k-1 \mid k-1), \mathbf{u}_i(k-1)\right) \qquad (4.95)$$

$$\mathbf{P}_i(k \mid k-1) = \nabla \mathbf{f}_{x_i}(k)\mathbf{P}_i(k-1 \mid k-1)\nabla \mathbf{f}_{x_i}{}^T(k) + \mathbf{Q}_i(k). \qquad (4.96)$$

The function $\mathbf{f}_i$ represents the local nonlinear state transition. Unlike in the DEKF, these predictions are different for each node. The state transition Jacobians are evaluated at different predicted local states and hence they are different between nodes. The state predictions are then used to generate local state estimates.

**Estimation**

The main linearized estimation and assimilation equations take the same form as those of the DDKF. The only difference is that the transformed and communicated information is dependent on nonlinear observations and models. The entire algorithm is presented here for completeness.

$$\hat{\mathbf{x}}_i(k \mid k) =$$

$$\mathbf{P}_i(k \mid k) \left\{ \mathbf{P}_i^+(k \mid k-1)\hat{\mathbf{x}}_i(k \mid k-1) + \sum_{j=1}^{N} \left[ \mathbf{P}_i^+(k \mid \mathbf{z}_j(k))\hat{\mathbf{x}}_i(k \mid \mathbf{z}_j(k)) \right] \right\}$$

$$\mathbf{P}_i(k \mid k) = \left[ \mathbf{P}_i^+(k \mid k-1) + \sum_{j=1}^{N} \mathbf{P}_i^+(k \mid \mathbf{z}_j(k)) \right]^+ .$$

The transformed covariance and state estimates are given by

$$\mathbf{P}_i(k \mid \mathbf{z}_j(k)) = \mathbf{T}_i(k) \left[ \mathbf{T}_j^T(k)\mathbf{P}_j^+(k \mid \mathbf{z}_j(k))\mathbf{T}_j(k) \right]^+ \mathbf{T}_i^T(k) \qquad (4.97)$$

$$\hat{\mathbf{x}}_i(k \mid \mathbf{z}_j(k)) = \mathbf{V}_{ji}(k)\hat{\mathbf{x}}_j(k \mid \mathbf{z}_j(k)). \qquad (4.98)$$

Local covariance and state estimates, based only on observations, at node $j$ are computed from local nonlinear observations and their models.

$$\mathbf{P}_j(k \mid \mathbf{z}_j(k)) = \mathbf{T}_j(k) \left[ \left\{ \nabla \mathbf{c}_{x_j}(k) \mathbf{T}_j(k) \right\}^T \mathbf{R}_j^+(k) \left\{ \nabla \mathbf{c}_{x_j}(k) \mathbf{T}_j(k) \right\} \right]^+ \mathbf{T}_j^T(k)$$

$$\hat{\mathbf{x}}_j(k \mid \mathbf{z}_j(k)) = \mathbf{P}_j(k \mid \mathbf{z}_j(k)) \mathbf{i}_j(\mathbf{z}_j(k)),$$

where the information contribution from observation $\mathbf{z}_j(k)$ is given by

$$\mathbf{i}_j(\mathbf{z}_j(k)) = \nabla \mathbf{c}_{x_j}^T(k) \mathbf{R}_j^+(k) \left[ \nu_j(k) + \nabla \mathbf{c}_{x_j}(k) \hat{\mathbf{x}}_j(k \mid k - 1) \right], \quad (4.99)$$

and $\mathbf{c}_j$ is the local nonlinear observation model such that

$$\mathbf{z}_j(k) = \mathbf{c}_j \left( \mathbf{x}_j(k), k \right). \quad (4.100)$$

Just as in the DDKF, the assimilation equations can be simplified by putting constraints on $\mathbf{T}_i(k)$, $\mathbf{T}_j(k)$ and $\mathbf{P}_j(k \mid \mathbf{z}_j(k))$. The DDEKF effectively extends the benefits of state space model distribution to nonlinear multisensor systems.

### 4.6.4 The Distributed and Decentralized Extended Information Filter (DDEIF)

Despite its ability to deal with and resolve nonlinear estimation problems, the DDEKF is still prone to the disadvantages of the DEKF. Communication in the DDEKF is not minimized. This is because the algorithm depends on $\mathbf{V}_{ji}(k)$ which satisfies the sufficient but not necessary observation model condition. In order to minimize communication, it is better to derive an information version of the DDEKF which uses $\mathbf{T}_{ji}(k)$. This is achieved by employing model distribution and internodal information transformation theory to the DEIF algorithm. The resulting algorithm is the distributed and decentralized extended information filter (**DDEIF**).

**Prediction**

The presence of local system models, which are different from global ones, distinguishes the DDEIF prediction equations from those of the DEIF. Other than this, the prediction procedures for the two filters are similar.

$$\hat{\mathbf{y}}_i(k \mid k - 1) = \mathbf{Y}_i(k \mid k - 1) \mathbf{f}_i \left( k, \hat{\mathbf{x}}_i(k - 1 \mid k - 1), \mathbf{u}_i(k - 1) \right) \quad (4.101)$$

$$\mathbf{Y}_i(k \mid k - 1) = \left[ \nabla \mathbf{f}_{x_i}(k) \mathbf{Y}_i^{-1}(k - 1 \mid k - 1) \nabla \mathbf{f}_{x_i}^T(k) + \mathbf{Q}_i(k) \right]^{-1}. \quad (4.102)$$

Unlike in the DEIF, these predictions are unique for each node. They are predictions of information about states locally relevant to node $i$. The predictions are then used in the computation of information estimates of the same states.

**Estimation**

The estimation equations are similar to those of the DEIF presented in Chapter 3. The linearized estimates depend on communicated transformed information from nonlinear observations taken at individual nodes. This information is assimilated to give reduced order estimates.

$$\hat{\mathbf{y}}_i(k \mid k) = \hat{\mathbf{y}}_i(k \mid k-1) + \sum_{j=1}^{N} \left[ \mathbf{T}_{ji}(k) \mathbf{i}_j(\mathbf{z}_j(k)) \right] \tag{4.103}$$

$$\mathbf{Y}_i(k \mid k) = \mathbf{Y}_i(k \mid k-1) + \sum_{j=1}^{N} \mathbf{I}_i(\mathbf{z}_j(k)). \tag{4.104}$$

The transformed associated information matrix and the information space internodal transformation are given by

$$\mathbf{I}_i(\mathbf{z}_j(k)) = \left[ \mathbf{T}_i(k) \left[ \mathbf{T}_j^T(k) \mathbf{I}_j(\mathbf{z}_j(k)) \mathbf{T}_j(k) \right]^+ \mathbf{T}_i^T(k) \right]^+ \tag{4.105}$$

$$\mathbf{T}_{ji}(k) = \mathbf{I}_i(\mathbf{z}_j(k)) \mathbf{V}_{ji}(k) \mathbf{I}_j^+(\mathbf{z}_j(k)). \tag{4.106}$$

The communicated information parameters are computed from local nonlinear observations and their linearized models.

$$\mathbf{I}_j(\mathbf{z}_j(k)) = \nabla \mathbf{c}_{x_j}^T(k) \mathbf{R}_j^+(k) \nabla \mathbf{c}_{x_j}(k) \tag{4.107}$$

$$\mathbf{i}_j(\mathbf{z}_j(k)) = \nabla \mathbf{c}_{x_j}^T(k) \mathbf{R}_j^+(k) \left[ \nu_j(k) + \nabla \mathbf{c}_{x_j}(k) \hat{\mathbf{x}}_j(k \mid k-1) \right]. \tag{4.108}$$

The vector $\nu_j(k)$ is the local nonlinear innovation,

$$\nu_j(k) = \mathbf{z}_j(k) - \mathbf{c}_j \left( \hat{\mathbf{x}}_j(k \mid k-1) \right). \tag{4.109}$$

As is the case with the DDIF, the estimation Equations 4.103 and 4.104 can be simplified by employing special cases of $\mathbf{T}_{ji}(k)$ outlined and discussed in Section 4.5. The DDEIF is the most generalized of all the decentralized estimation algorithms presented in this book. It integrates the advantages of nonlinear information space with the benefits of both decentralization and model distribution. It has potential applications in nonlinear multisensor estimation problems. Improved scalability, reduced computation and minimized communication are its advantages over the DEIF and DDEKF. However, like the DEKF, DEIF and DDEKF, the DDEIF algorithm is vulnerable to linearization instability.

## 4.7 Summary

In this chapter scalable decentralized estimation methods have been derived for both linear and nonlinear systems. Both information space and state space algorithms were derived and compared. These algorithms are based on generalized internodal communication and transformation theory also developed in this chapter. This theory allows communication between nodes to be minimized, both in terms of the number of communication links and size of message. The nodal transformations are model defined; that is, they depend on the system state transition model. The key novelty of the resulting non-fully connected topologies is that there is no need to propagate information between two unconnected nodes. This makes the algorithms simpler, more efficient and scalable than previous methods. The estimation algorithms presented provide exactly the same results as those obtained from an equivalent fully connected system or conventional centralized system.

# Chapter 5

## Scalable Decentralized Control

### 5.1 Introduction

This chapter extends the decentralized estimation algorithms of Chapters 3 and 4 to the problem of sensor based control. The chapter starts by introducing stochastic control ideas. In particular, the LQG control problem and its solution are outlined. For systems involving nonlinearities, the nonlinear stochastic control problem is discussed. Principles of stochastic control are then extended to multisensor and multiactuator systems. This is done by using decentralized estimation methods based on the DIF and DEIF algorithms. The advantages and limitations of both centralized and *fully connected* decentralized control systems are discussed. A distributed and decentralized control algorithm is proposed as a solution to the limitations of fully connected decentralized control systems. This is done by extending the scalable decentralized estimation algorithms, the DDIF and DDEIF to control systems. The resulting *modular* and *scalable* control topology is the major *composite* theoretical result of this book. Its operation principles and attributes are described. The models of the simulation example used to validate the theory are then presented.

### 5.2 Optimal Stochastic Control

This section describes the *optimal stochastic control* problem and its solution. The practical design of stochastic controllers for problems described by the **LQG** assumptions, *Linear* system model, *Quadratic* cost criterion for optimality, and *Gaussian* white noise inputs are briefly discussed. Problems involving nonlinear models are then considered.

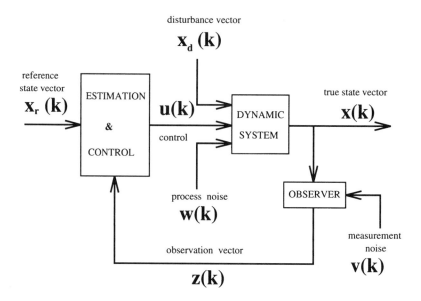

**FIGURE 5.1**
**Stochastic Control System Configuration**

## 5.2.1   Stochastic Control Problem

Most control problems of interest can be described by the general system configuration in Figure 5.1. There is some *dynamic system* of interest, whose behavior is to be affected by an applied *control* input vector $\mathbf{u}(k)$ in such a way that the *controlled* state vector $\mathbf{x}(k)$ exhibits desirable characteristics. These characteristics are prescribed, in part, as the controlled state vector $\mathbf{x}(k)$ matching a *reference* state vector $\mathbf{x_r}(k)$ as closely and quickly as possible. The simplest control problem is in the LQG form and hence it is important to understand the meaning of the LQG assumptions.

- **Linear System Model**: Linearity is assumed where a linear system obeys the principle of superposition and its response is the convolution of the input with the system impulse response.

- **Quadratic Cost Function**: A quadratic cost criterion for optimality is assumed such that the control is optimal in the sense of minimizing the expected value of a quadratic performance index associated with the control problem.

- **Gaussian Noise Model**: White Gaussian noise process corruption is assumed.

**Problem Statement**

Let the system of interest be described by the $n$-dimensional stochastic discrete time difference equation

$$\mathbf{x}(k) = \mathbf{F}(k)\mathbf{x}(k-1) + \mathbf{B}(k)\mathbf{u}(k-1) + \mathbf{D}(k)\mathbf{w}(k), \qquad (5.1)$$

where $\mathbf{u}(k)$ is the $r$-dimensional control input to be applied and $\mathbf{w}(k)$ is the zero mean white Gaussian discrete time noise. The objective is to determine the control vector $\mathbf{u}(k)$ that minimizes the quadratic cost function

$$\mathbf{J}(N) = \mathrm{E}\left[\sum_{i=1}^{N} \left\{ \mathbf{e}_r^T(k)\mathbf{X}(k)\mathbf{e}_r(k) + [\mathbf{u}(k) - \mathbf{u}_r(k)]^T\mathbf{U}(k)[\mathbf{u}(k) - \mathbf{u}_r(k)] \right\} \right],$$

where $\mathbf{e}_r(k) = [\mathbf{x}(k) - \mathbf{x}_r(k)]$ and $\mathbf{X}(k)$ is an $n$-by-$n$ real and positive semi-definite cost weighting matrix, reflecting the relative importance of maintaining individual state component deviations at small values. $\mathbf{U}(k)$ is an $r$-by-$r$ real, symmetric and positive definite cost weighting matrix reflecting the relative importance of maintaining individual control component deviations at small values [76].

There are several reasons for the use of a quadratic cost function of states and control:

- Quadratics are a good description of many control objectives, such as minimizing mean squared error or energy.

- Inherently such a function enhances the adequacy of the linear perturbation model.

- This combination of modeling assumptions yields a *tractable* problem whose solution is in the form of a readily synthesized, efficiently implemented, feedback control law.

## 5.2.2 Optimal Stochastic Solution

In this subsection the solution to the LQG control problem outlined above is presented. Deterministic methods cannot be used to solve for an optimal control vector $\mathbf{u}(k)$ from the function $\mathbf{J}(N)$ because of the stochastic nature of the problem [76]. The dynamic driving noise term $\mathbf{w}(k)$ prevents perfect, ahead-of-time knowledge of where the system will be at time $(k+1)$. There is no single optimal history of states and controls, but an entire family of trajectories. Two closely related techniques are employed in determining an optimal stochastic control solution [76].

- **Optimality principle**: An optimal policy has the property that for any initial states and decision *(control law)*, all remaining decisions must constitute an optimal policy with regard to the state which results from the first decision.

- **Stochastic dynamic programming**: This is a technique of step-ping backward in time to obtain optimal control. It is dependent on the *Markov* nature of the discrete-time process.

Two further structural properties are essential for the solution to be realized. These are *separation* and *certainty equivalence principles*. A control problem is said to be separable if its optimal control depends only on an estimate $\hat{\mathbf{x}}(k \mid k)$ of the state $\mathbf{x}(k)$ and not at all on the accuracy of the estimate. It is also said to be certainty equivalent if, being separable, the control is exactly the same as it would be in a related deterministic problem. The two principles imply that the problem of seeking a linear control law for a linear dynamical system with Gaussian measurement noise subject to a quadratic performance index can be cast in terms of two separate problems:

- Optimal deterministic control

- Optimal stochastic estimation

These two problems can be solved separately to yield an optimal solution to the combined problem. The optimal stochastic estimation problem has been solved in Chapter 2 for single sensor systems and in Chapters 3 and 4 for multisensor systems. The basis of these algorithms is the Kalman filter and its algebraic equivalent, the Information filter. Although only the information space algorithms are extended to stochastic control algorithms in this chapter, the state space estimation algorithms can be similarly extended.

The cost minimizing control function is given by

$$\mathbf{u}(k) = -\mathbf{G}(k)[\hat{\mathbf{x}}(k \mid k) - \mathbf{x}_r(k)], \tag{5.2}$$

where $\mathbf{G}(k)$ is the associated optimal deterministic control gain. Its value is generated from the solution to the *Backward Riccati recursion* [76],

$$\mathbf{G}(k) = [\mathbf{U}(k) + \mathbf{B}^T(k)\mathbf{K}(k)\mathbf{B}(k)]^{-1}[\mathbf{B}^T(k)\mathbf{K}(k)\mathbf{F}(k)], \tag{5.3}$$

where $\mathbf{K}(k)$ is the $n$-by-$n$ symmetric matrix satisfying the *Backward Riccati difference* equation [76],

$$\begin{aligned}
\mathbf{K}(k) &= \mathbf{X}(k) + \mathbf{F}^T(k)\mathbf{K}(k+1)\mathbf{F}(k) - \left[\mathbf{F}^T(k)\mathbf{K}(k+1)\mathbf{B}(k)\mathbf{G}(k)\right] \\
&= \mathbf{X}(k) + \left[\mathbf{F}^T(k)\mathbf{K}(k+1)\right]\left[\mathbf{F}(k) - \mathbf{B}(k)\mathbf{G}(k)\right]. \tag{5.4}
\end{aligned}$$

This equation is solved backwards from the terminal condition, $\mathbf{K}(N+1) = \mathbf{X}_f(k)$. The *untracked* state estimate $\hat{\mathbf{x}}(k \mid k)$ is reconstructed from the *tracked* information estimate and the (information matrix),

$$\hat{\mathbf{x}}(k \mid k) = \mathbf{Y}^{-1}(k \mid k)\hat{\mathbf{y}}(k \mid k). \tag{5.5}$$

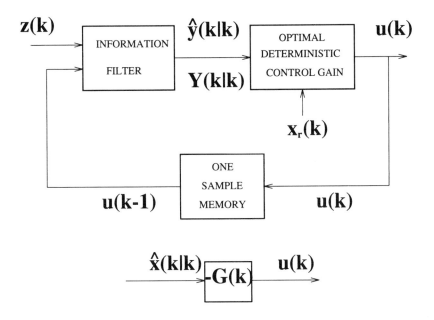

**FIGURE 5.2**
**Optimal Stochastic Control**

**Solution Statement**

The optimal stochastic control for a problem described by linear system models driven by white Gaussian noise, subject to a quadratic cost criterion, consists of an optimal linear Information filter cascaded with the optimal feedback gain matrix of the corresponding deterministic optimal control problem. This means the optimal stochastic control function is equivalent to the associated optimal deterministic control function with the true state replaced by the conditional mean of the state given the measurements. Illustration of this stochastic control solution is shown in Figure 5.2.

The importance of this result is the synthesis capability it yields. Under the LQG assumptions, the design of the optimal stochastic controller can be completely separated into the design of the appropriate information filter and the design of an optimal deterministic controller associated with the original problem. The feedback control gain matrix is *independent* of all uncertainty, so a controller can be designed assuming that $\mathbf{x}(k)$ is known perfectly all the time. Similarly, the filter is *independent* of the matrices that define the controller performance measures. The estimation algorithm can thus be developed ignoring the fact that a control problem is under consideration.

**Algorithm Summary**

Estimation is carried out according to the Information filter Equations 2.37 and 2.38. The information estimate $\hat{\mathbf{y}}(k \mid k)$ is used to generate the state estimate and then the control signal.

$$\hat{\mathbf{x}}(k \mid k) = \mathbf{Y}^{-1}(k \mid k)\hat{\mathbf{y}}(k \mid k) \tag{5.6}$$

$$\mathbf{u}(k) = -\mathbf{G}(k)\left[\hat{\mathbf{x}}(k \mid k) - \hat{\mathbf{x}}_r(k \mid k)\right]. \tag{5.7}$$

The control law is generated as follows:

$$\mathbf{G}(k) = \left[\mathbf{U}(k) + \mathbf{B}^T(k)\mathbf{K}(k)\mathbf{B}(k)\right]^{-1}\left[\mathbf{B}^T(k)\mathbf{K}(k)\mathbf{F}(k)\right] \tag{5.8}$$

$$\mathbf{K}(k) = \mathbf{X}(k) + \left[\mathbf{F}^T(k)\mathbf{K}(k+1)\right]\left[\mathbf{F}(k) - \mathbf{B}(k)\mathbf{G}(k)\right]. \tag{5.9}$$

This is the optimal stochastic LQG control solution for single sensor and single actuator system. Before extending it to multisensor and multiactuator systems, the case of stochastic control problems with nonlinearities is considered.

## 5.2.3    Nonlinear Stochastic Control

The *separation* and *certainty* equivalence principles do not hold for nonlinear systems. Several methods have been employed in literature to attempt to solve this problem [76]. These include *linear perturbation control* (LQG direct synthesis), *closed-loop controller* ("dual control" approximation) and *stochastic adaptive control.* In this book *assumed certainty equivalence* design is used. This is a synthesis technique which *separates* the stochastic controller into the cascade of an estimator and a deterministic optimal control function even when the *optimal* stochastic controller does *not* have the certainty equivalence property. It must be emphasized that, by definition, certainty equivalence assumes that the separation principle holds. Thus, the first objective is to solve the associated deterministic optimal control, ignoring the uncertainties and assuming perfect access to the entire state. Deterministic dynamic programming is used to generate the control law as a feedback law. The second objective is to solve the *nonlinear estimation* problem. This has already been done in Chapter 2, by deriving the EKF and EIF. In order to utilize the advantages of information space, the EIF is used.

Finally, the assumed certainty equivalence control law is computed by substituting the *linearized* information estimate from the EIF in the deterministic control law. This is the  assumed certainty equivalence nonlinear stochastic control algorithm, illustrated in Figure 5.3. One important special case of this design methodology is the cascading of an EIF equivalent of a *constant gain* EKF to a *constant gain* linear quadratic state feedback

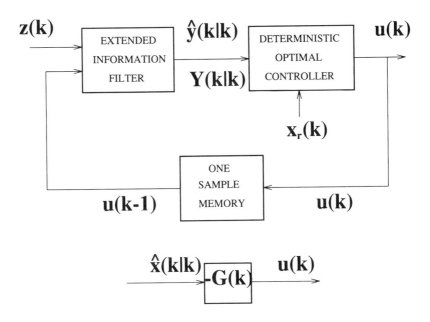

**FIGURE 5.3**
**Nonlinear Stochastic Control**

controller. The constant gain EKF has the basic structure of an EKF, except that the constant gain is precomputed based on linearization about the *nominal* trajectory. This filter is robust against divergence. However, there is no fundamental reason to limit attention to constant gain designs other than computational burden of the resulting algorithms.

Equipped with both *single* sensor *LQG* and nonlinear stochastic control algorithms, the next step is to extend them to multisensor and multiactuator control systems.

### 5.2.4  Centralized Control

Just as in the multisensor data fusion case, there are three broad categories of multiactuator control architectures: centralized, hierarchical and decentralized. A centralized control system consists of multiple sensors forming a *decentralized observer* outlined in Chapter 3. The control realization remains centrally placed, whereby information from the sensors is globally fused to generate a control law. Figure 5.4 shows such a system. Only observations are locally taken and sent to a center where estimation and control occurs centrally.

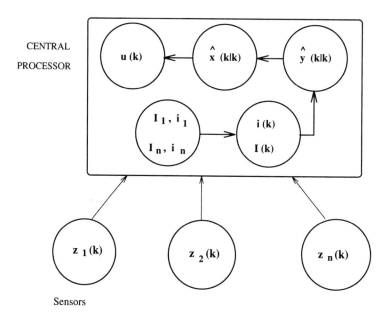

**FIGURE 5.4**
**Centralized Control**

The information prediction equations are the same as those of a single sensor system, Equations 2.35 and 2.36.

**Control Generation**

Global information estimates are centrally computed from global information predictions and observations generated by the different sensors. The state estimate is reconstructed from the tracked central information vector and matrix. The control vector is then computed from the state error and globally generated control law. The entire algorithm is illustrated in Figure 5.4. The main feature of this arrangement is the ability to employ several sensors, while retaining a single, central actuator.

## 5.3 Decentralized Multisensor Based Control

In addition to using multiple sensors, it would be even more beneficial if multiple actuators could be used, such that control achieved locally is the same as that achieved with a centralized controller. This is the motivation behind decentralized multisensor based control. The approach adopted is to initially derive a *fully connected* decentralized control system, and then proceed to eliminate the full connection constraint to produce a scalable decentralized control system.

## 5.3.1 Fully Connected Decentralized Control

The derivation of a fully connected control network is essentially an extension of the decentralized Information filter, DIF, to a decentralized information form of the standard LQG controller. This system consists of fully connected network of communicating control nodes. Each control node has a local Information filter, communicates with other nodes and then generates a global optimal control vector. Figure 5.5 illustrates a typical fully connected control configuration of four nodes. An algorithm is developed for an arbitrary number of nodes. The communication and estimation equations are as in the DIF algorithm.

**Control Generation**

Control equations for each actuator are obtained from those of the single actuator (centralized) system. Since the system is fully connected, the nodal control models are the same as those in the centralized system. The global state estimate is computed locally.

$$\hat{x}_i(k \mid k) = Y_i^{-1}(k \mid k)\hat{y}_i(k \mid k). \tag{5.10}$$

A local control law is then obtained with respect to a local reference vector

$$u_i(k) = -G_i(k)\left[\hat{x}_i(k \mid k) - x_{r_i}(k)\right], \tag{5.11}$$

where the nodal control gain is obtained from

$$G_i(k) = \left[U(k) + B^T(k)K_i(k)B(k)\right]^{-1}\left[B^T(k)K_i(k)F(k)\right] \tag{5.12}$$

$$K_i(k) = X(k) + \left[F^T(k)K_i(k+1)\right]\left[F(k) - B(k)G_i(k)\right]. \tag{5.13}$$

$X(k)$ is a state cost weighting matrix and $U(k)$ is a control cost weighting matrix. $K_i(k)$ is the decentralized Backward Riccati difference matrix. This derivation assumes that each local control node has a state space model and information space identical to the corresponding centralized (global) descriptions. Hence, the sizes of the local control models are the same as in the centralized system.

For the decentralized control network to be equivalent to the centralized one, all the nodes must communicate. Initialization of information vectors and matrices must be the same at each node. Figure 5.5 shows a fully connected network of four control nodes. From this diagram it is evident that

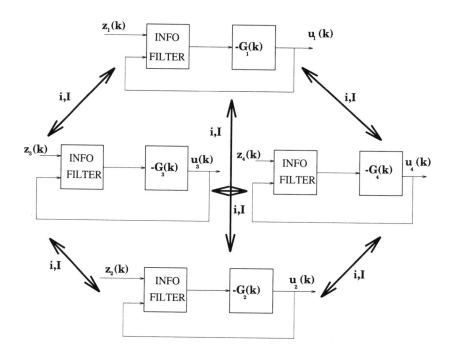

**FIGURE 5.5**
**Fully Connected Decentralized Control**

each node consists of an information form of the standard LQG controller. This controller takes local observation, communicates with all the other three nodes and locally generates the global control vector. The control vector obtained by each node is exactly the same for all four nodes and also identical to that generated in a conventional centralized LQG controller.

Similarly, the nonlinear decentralized control algorithm is obtained by employing the principles of nonlinear stochastic control. Instead of the Information filter, the extended Information filter (EIF) is used at each node. The prediction and estimation equations are the same as those of the DEIF. Employing assumed certainty equivalence, the decentralized deterministic control generation is the same as in the linear case given in Equations 5.10 to 5.13.

### The Drawbacks

A fully connected decentralized control system has the benefits of decentralization presented and discussed in Chapter 3. These include modularity, robustness and flexibility of control nodes. However, the algorithm also manifests the problems associated with fully connected decentralization: limited scalability, excessive computation and extensive communication. These drawbacks are discussed in detail in Chapter 3. In particular, the local control vector $\mathbf{u}_i(k)$ is the same size as the centralized control vector $\mathbf{u}(k)$. Similarly, the sizes of control models are not reduced. The network replicates the central (global) control at each node and is characterized by limited scalability. Applications are limited for such a computationally redundant system.

## 5.3.2 Distribution of Control Models

The limitations of fully connected decentralized control systems can be removed by using model distribution and local internodal communication. The principle is to eradicate the problems while retaining the benefits of decentralization. In Chapter 4 model distribution is used to derive two scalable estimation algorithms, the DDIF and DDEIF. The objective here is to extend these algorithms to *distributed* and decentralized control systems.

The starting point is distributing the control models. The local state transition equation is given by

$$\mathbf{x}_i(k+1) = \mathbf{F}_i(k)\mathbf{x}_i(k) + \mathbf{B}_i(k)\mathbf{u}_i(k) + \mathbf{D}_i(k)\mathbf{w}_i(k), \tag{5.14}$$

where $\mathbf{B}_i(k)$ and $\mathbf{u}_i(k)$ are the local control gain matrix and control vector, respectively. They are generally different from the those in the global system. However, the local states are controlled in the same way that they would be controlled in a centralized or fully connected decentralized control system. The local state transition and noise models are obtained as derived in Chapter 4.

$$\mathbf{F}_i(k) = \mathbf{T}_i(k)\mathbf{F}(k)\mathbf{T}_i^+(k-1) \tag{5.15}$$
$$\mathbf{D}_i(k) = \mathbf{T}_i(k)\mathbf{D}(k)\mathbf{T}_i^+(k-1). \tag{5.16}$$

The local control models are similarly derived such that local estimates and control are exactly the same as in the global system [113], [20]. The new local *state transition* equation is given by

$$
\begin{aligned}
\mathbf{x}_i(k) &= \mathbf{F}_i(k)\mathbf{x}_i(k-1) + \mathbf{B}_i(k)\mathbf{u}_i(k-1) + \mathbf{D}_i(k)\mathbf{w}_i(k-1) \quad (5.17)\\
&= \mathbf{F}_i(k)\mathbf{T}_i(k-1)\mathbf{x}(k-1) + \mathbf{B}_i(k)\mathbf{G}_i(k)\mathbf{T}_i(k-1) \times \\
&\quad [\mathbf{x}(k-1) - \mathbf{x_r}(k-1)] + \mathbf{D}_i(k)\mathbf{T}_i(k-1)\mathbf{w}(k-1) \\
\Leftrightarrow \mathbf{T}_i(k)\mathbf{x}(k) &= \mathbf{F}_i(k)\mathbf{T}_i(k-1)\mathbf{x}(k-1) + \mathbf{B}_i(k)\mathbf{G}_i(k)\mathbf{T}_i(k-1) \times \\
&\quad [\mathbf{x}(k-1) - \mathbf{x_r}(k-1)] + \mathbf{D}_i(k)\mathbf{T}_i(k-1)\mathbf{w}(k-1) \quad (5.18)
\end{aligned}
$$

where $\mathbf{x_r}(k-1)$ is the previous reference state vector.

Pre-multiplying the global state transition Equation 2.1 by $\mathbf{T}_i(k)$ through-out gives

$$\mathbf{T}_i(k) \times \mathbf{x}(k)$$
$$= \mathbf{T}_i(k)\mathbf{F}(k)\mathbf{x}(k-1) + \mathbf{T}_i(k)\mathbf{B}(k)\mathbf{u}(k-1) + \mathbf{T}_i(k)\mathbf{D}(k)\mathbf{w}(k-1)$$
$$= \mathbf{T}_i(k)\mathbf{F}(k)\mathbf{x}(k-1) + \mathbf{T}_i(k)\mathbf{B}(k)\mathbf{G}(k)[\mathbf{x}(k-1) - \mathbf{x_r}(k-1)] +$$
$$\mathbf{T}_i(k)\mathbf{D}(k)\mathbf{w}(k-1). \tag{5.19}$$

Comparing Equations 5.18 and 5.19 and equating the coefficients of the state error signal $[\mathbf{x}(k-1) - \mathbf{x_r}(k-1)]$ give

$$[\mathbf{B}_i(k)\mathbf{G}_i(k)]\,\mathbf{T}_i(k-1) = \mathbf{T}_i(k)[\mathbf{B}(k)\mathbf{G}(k)]. \tag{5.20}$$

The products $[\mathbf{B}_i(k)\mathbf{G}_i(k)]$ and $[\mathbf{B}(k)\mathbf{G}(k)]$ are the *effective* nodal and global control gains applied to the global and nodal state error vectors, respectively. Equation 5.20 is true for any nodal transformation. It is a discrete time, control version of Sandell's continuous time (necessary and sufficient) dynamic *equivalence* condition [112]. For a nodal transformation matrix, $\mathbf{T}_i(k-1) \in \mathcal{R}^{m \times n}$, which produces model size reduction with no redundant states in the local state vector, it follows that $m \leq n$ and rank $\mathbf{T}_i(k-1) = m$. This means that $\mathbf{T}_i(k-1)$ is full row rank and hence it has a *right* inverse. This allows the extraction of an expression for the nodal control models from the discrete time Sandell's condition

$$\mathbf{B}_i(k)\mathbf{G}_i(k) = \mathbf{T}_i(k)[\mathbf{B}(k)\mathbf{G}(k)]\mathbf{T}_i^+(k-1). \tag{5.21}$$

### 5.3.3   Distributed and Decentralized Control

The objective here is to derive a nodal control algorithm which is completely expressed in terms of information. Although the local control vector is unique for each different node, the control signal for each locally relevant state should be the same as that obtained in a centralized state space system. The local state estimate is reconstructed from the information estimates.

$$\hat{\mathbf{x}}_i(k \mid k) = \mathbf{Y}_i^{-1}(k \mid k)\hat{\mathbf{y}}_i(k \mid k). \tag{5.22}$$

Unlike in fully connected decentralized control problems, this estimate is different for different nodes. The estimate is used to compute a local state space control vector with respect to a local reference vector. The control vector is then expressed in terms of information.

$$\mathbf{u}_i(k)$$
$$= -\mathbf{G}_i(k)\left[\hat{\mathbf{x}}_i(k \mid k) - \mathbf{x}_{r_i}(k)\right]$$
$$= -\left[\mathbf{B}_i^+(k)\mathbf{B}_i(k)\right]\mathbf{G}_i(k)\left[\hat{\mathbf{x}}_i(k \mid k) - \mathbf{x}_{r_i}(k)\right] \tag{5.23}$$

$$= -\mathbf{B}_i^+(k) \left[ \mathbf{B}_i(k)\mathbf{G}_i(k) \right] \left[ \hat{\mathbf{x}}_i(k \mid k) - \mathbf{x}_{r_i}(k) \right]$$
$$= -\mathbf{B}_i^+(k)\mathbf{T}_i(k+1) \left[ \mathbf{B}(k)\mathbf{G}(k) \right] \mathbf{T}_i^+(k) \left[ \hat{\mathbf{x}}_i(k \mid k) - \mathbf{x}_{r_i}(k) \right]$$
$$= -\mathbf{B}_i^+(k)\mathbf{T}_i(k+1) \left[ \mathbf{B}(k)\mathbf{G}(k) \right] \mathbf{T}_i^+(k)\mathbf{Y}_i^{-1}(k \mid k) \left[ \hat{\mathbf{y}}_i(k \mid k) - \mathbf{y}_{r_i}(k) \right]$$
$$\triangleq -\mathbf{\Omega}_i(k \mid k)\delta_i(k). \tag{5.24}$$

This is the *distributed* and decentralized information control law generated by each node where the local information error $\delta_i(k)$ and local control information gain $\mathbf{\Omega}_i(k \mid k)$ are given by

$$\delta_i(k) \triangleq \hat{\mathbf{y}}_i(k \mid k) - \mathbf{y}_{r_i}(k) \tag{5.25}$$

$$\mathbf{\Omega}_i(k \mid k) \triangleq \left\{ \mathbf{B}_i^+(k)\mathbf{T}_i(k+1) \left[ \mathbf{B}(k)\mathbf{G}(k) \right] \mathbf{T}_i^+(k) \right\} \mathbf{Y}_i^{-1}(k \mid k)$$
$$= \mathbf{G}_i(k)\mathbf{Y}_i^{-1}(k \mid k). \tag{5.26}$$

The vector $\mathbf{y}_{r_i}(k)$ is the local reference information state vector obtained from the reference state and information matrix estimate,

$$\mathbf{y}_{r_i}(k) = \mathbf{Y}_i(k \mid k)\mathbf{x}_{r_i}(k). \tag{5.27}$$

$\mathbf{G}_i(k)$ and $\mathbf{B}_i(k)\mathbf{G}_i(k)$ are the corresponding local state space control and *effective* control gains, respectively. They are obtained as follows:

$$\mathbf{G}_i(k) \triangleq \mathbf{B}^+{}_i(k)\mathbf{T}_i(k+1) \left[ \mathbf{B}(k)\mathbf{G}(k) \right] \mathbf{T}_i^+(k) \tag{5.28}$$

$$\Leftrightarrow \mathbf{B}_i(k)\mathbf{G}_i(k) \triangleq \mathbf{T}_i(k+1) \left[ \mathbf{B}(k)\mathbf{G}(k) \right] \mathbf{T}_i^+(k). \tag{5.29}$$

The computation of $\mathbf{G}_i(k)$ is carried out from local control models using distributed and decentralized dynamic programming.

$$\mathbf{G}_i(k) = \left[ \mathbf{U}_i(k) + \mathbf{B}_i^T(k)\mathbf{K}_i(k)\mathbf{B}_i(k) \right]^{-1} \left[ \mathbf{B}_i^T(k)\mathbf{K}_i(k)\mathbf{F}_i(k) \right],$$

where

$$\mathbf{B}_i(k) = \left\{ \mathbf{T}_i(k+1) \left[ \mathbf{B}(k)\mathbf{G}(k) \right] \mathbf{T}_i^+(k) \right\} \mathbf{G}_i^+(k)$$
$$\mathbf{G}(k) = \mathbf{B}^+(k) \left\{ \mathbf{T}_i^+(k+1) \left[ \mathbf{B}_i(k)\mathbf{G}_i(k) \right] \mathbf{T}_i(k) \right\}$$
$$\mathbf{K}_i(k) = \mathbf{X}_i(k) + \left[ \mathbf{F}_i^T(k)\mathbf{K}_i(k+1) \right] \left[ \mathbf{F}_i(k) - \mathbf{B}_i(k)\mathbf{G}_i(k) \right].$$

$\mathbf{X}_i(k)$ is a *local* state cost weighting matrix and $\mathbf{U}_i(k)$ a *local* control cost weighting matrix. $\mathbf{K}_i(k)$ is the distributed and decentralized Backward Riccati difference matrix. All computation is carried out locally and $\mathbf{B}(k)$ is available locally. If $\mathbf{B}(k)$ is equal to the identity matrix $\mathbf{1}$, that is, all states are directly controlled, then the computation simplifies to

$$\mathbf{B}_i(k) = \mathbf{1}$$
$$\mathbf{G}_i(k) = \left[ \mathbf{U}_i(k) + \mathbf{K}_i(k) \right]^{-1} \left[ \mathbf{K}_i(k)\mathbf{F}_i(k) \right]$$
$$\mathbf{K}_i(k) = \mathbf{X}_i(k) + \left[ \mathbf{F}_i^T(k)\mathbf{K}_i(k+1) \right] \left[ \mathbf{F}_i(k) - \mathbf{G}_i(k) \right].$$

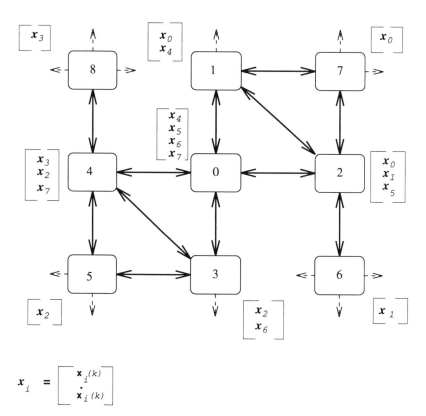

FIGURE 5.6
Scalable Control Network

### 5.3.4  System Characteristics

Several features of this distributed and decentralized control system distinguish it from centralized and *fully connected* decentralized control systems. The algorithm retains all the advantages of full decentralization while resolving the problems of full connectedness. In general, the characteristics of this scalable control system are independently shared by its constituent estimation algorithms, the DDIF and the DDEIF.

- **Computation:** Reduced order models, vectors and matrices are used at each node and only relevant local computation is carried out. The optimal control gain, control law and Backward Riccati difference equation are partitioned into locally relevant components. Consequently, the memory required and computational load are reduced.

- **Communication:** This depends on connectedness, which in turn depends on internodal transformations. Two nodes $i$ and $j$ will communicate if, and only if, they have an overlapping information space, that is, at least one of the internodal transformation matrices, $\mathbf{T}_{ij}(k)$ and $\mathbf{T}_{ji}(k)$ is not a null matrix. When communication does take place, only relevant information is exchanged.

- **Topology:** The system is not necessarily fully connected. Arbitrary tree, ring or loop connected topologies based on internodal transformations are possible. A typical topology is shown in Figure 5.6. In fact, the fully connected decentralized, the hierarchical and the centralized control configurations are special cases of the distributed decentralized control network where for all nodes: $\{\mathbf{T}_{ij}(k) = \mathbf{1}\}$, $\{\mathbf{T}_{ij}(k) = \mathbf{T}_{iG}(k)\}$ and $\{\mathbf{T}_{ij}(k) = \mathbf{0}\}$, respectively. The model and control distribution is neither random nor *ad hoc*; it is dependent on $\mathbf{T}_{ij}(k)$. This creates scope for maximizing the benefits of distribution, while minimizing redundancy.

## 5.4  Simulation Example

The coupled mass system in Figure 4.1, discussed as an extended example in chapter 4, is used in simulation to demonstrate decentralization, model distribution and internodal transformation. The details of the models used are presented in this section. The simulation results are presented and discussed in Chapter 7.

### 5.4.1  Continuous Time Models

Each of the four masses is conceptually decoupled and the forces operating on it analyzed. As a result the following equations are derived:

$$\ddot{x}_1 + (b/m_1)\dot{x}_1 + (k/m_1)(x_1 - x_2) = (u_1/m_1)$$
$$\ddot{x}_2 + (b/m_2)\dot{x}_2 + (k/m_2)(2x_2 - x_1 - x_3) = (u_2/m_2)$$
$$\ddot{x}_3 + (b/m_3)\dot{x}_3 + (k/m_3)(2x_3 - x_2 - x_4) = (u_3/m_3)$$
$$\ddot{x}_4 + (b/m_4)\dot{x}_4 + (k/m_4)(x_4 - x_3) = (u_4/m_4).$$

Rearranging the free body equations,

$$\dot{x}_1 = x_5$$
$$\dot{x}_2 = x_6$$
$$\dot{x}_3 = x_7$$

$$\dot{x}_4 = x_8$$
$$\dot{x}_5 = -(k/m_1)x_1 + (k/m_1)x_2 - (b/m_1)x_5$$
$$\dot{x}_6 = (k/m_2)x_1 - 2(k/m_2)x_2 + (k/m_2)x_3 - (b/m_2)x_6$$
$$\dot{x}_7 = (k/m_3)x_2 - 2(k/m_3)x_3 + (k/m_3)x_4 - (b/m_3)x_7$$
$$\dot{x}_8 = (k/m_4)x_3 - (k/m_4x_4 - (b/m_4)x_8.$$

The continuous time system models in the equation

$$\dot{\mathbf{x}}(t) = \mathbf{A}\mathbf{x}(t) + \mathbf{B}\mathbf{u}(t) + \mathbf{w}(t) \qquad (5.30)$$

are then given by

$$
\mathbf{A} = \begin{bmatrix}
0 & 0 & 0 & 0 & 1 & 0 & 0 & 0 \\
0 & 0 & 0 & 0 & 0 & 1 & 0 & 0 \\
0 & 0 & 0 & 0 & 0 & 0 & 1 & 0 \\
0 & 0 & 0 & 0 & 0 & 0 & 0 & 1 \\
-k/m_1 & k/m_1 & 0 & 0 & -b/m_1 & 0 & 0 & 0 \\
k/m_2 & -2k/m_2 & k/m_2 & 0 & 0 & -b/m_2 & 0 & 0 \\
0 & k/m_3 & -2k/m_3 & k/m_3 & 0 & 0 & -b/m_3 & 0 \\
0 & 0 & k/m_4 & -k/m_4 & 0 & 0 & 0 & -b/m_4
\end{bmatrix}
$$

$$
\mathbf{B} = \begin{bmatrix}
0 & 0 & 0 & 0 \\
0 & 0 & 0 & 0 \\
0 & 0 & 0 & 0 \\
0 & 0 & 0 & 0 \\
(1/m_1) & 0 & 0 & 0 \\
0 & (1/m_2) & 0 & 0 \\
0 & 0 & (1/m_3) & 0 \\
0 & 0 & 0 & (1/m_4)
\end{bmatrix}.
$$

Using the following numerical data,

$$k_1 = k_2 = k_3 = k_4 = 50 N/m$$
$$m_1 = m_2 = m_3 = m_4 = 1.0 Kg$$
$$b_1 = b_2 = b_3 = b_4 = 0.1 N/(m/s)$$
$$u_1 = u_2 = u_3 = u_4 = 10 N,$$

gives the following models:

$$
\mathbf{A} = \begin{bmatrix}
0 & 0 & 0 & 0 & 1 & 0 & 0 & 0 \\
0 & 0 & 0 & 0 & 0 & 1 & 0 & 0 \\
0 & 0 & 0 & 0 & 0 & 0 & 1 & 0 \\
0 & 0 & 0 & 0 & 0 & 0 & 0 & 1 \\
-50 & 50 & 0 & 0 & -0.1 & 0 & 0 & 0 \\
50 & -100 & 50 & 0 & 0 & -0.1 & 0 & 0 \\
0 & 50 & -100 & 50 & 0 & 0 & -0.1 & 0 \\
0 & 0 & 50 & -50 & 0 & 0 & 0 & -0.1
\end{bmatrix}
$$

$$
\mathbf{B} = \begin{bmatrix}
0 & 0 & 0 & 0 \\
0 & 0 & 0 & 0 \\
0 & 0 & 0 & 0 \\
0 & 0 & 0 & 0 \\
1 & 0 & 0 & 0 \\
0 & 1 & 0 & 0 \\
0 & 0 & 1 & 0 \\
0 & 0 & 0 & 1
\end{bmatrix}.
$$

## 5.4.2 Discrete Time Global Models

The state transition matrix $\mathbf{F}(k)$ and the input control matrix $\mathbf{B}(k)$ are derived from the continuous time model matrices $\mathbf{A}$ and $\mathbf{B}$. The state transition matrix $\mathbf{F}(k)$ is computed by the Series method where for linear time-invariant systems,

$$
\mathbf{F}(k) = e^{\mathbf{A}(\Delta T)} = \mathbf{I} + \sum_{i=1}^{n} \{(\Delta T)^i \mathbf{A}^i\}/i!
$$

$$
\mathbf{B}(k) = \int_{k}^{k+1} Be^{\mathbf{A}\{(k+1)-\Delta T\}}.
$$

A discrete time approximation can be applied if $\Delta T$ is sufficiently small compared with the *time constants* of the system.

$$
\mathbf{F}(k) = \mathbf{I} + (\Delta T)\mathbf{A}
$$
$$
\mathbf{B}(k) = (\Delta T)\mathbf{B}.
$$

For the mass system both the approximation and the general method give the same results. This is because $\Delta T$, which was taken as 1.0 sec, is sufficiently small compared to the time constants of the system. The following system and observation models are obtained:

$$\mathbf{F}(k) = \begin{bmatrix} 1 & 0 & 0 & 0 & 1 & 0 & 0 & 0 \\ 0 & 1 & 0 & 0 & 0 & 1 & 0 & 0 \\ 0 & 0 & 1 & 0 & 0 & 0 & 1 & 0 \\ 0 & 0 & 0 & 1 & 0 & 0 & 0 & 1 \\ -50 & 50 & 0 & 0 & 0.9 & 0 & 0 & 0 \\ 50 & -100 & 50 & 0 & 0 & 0.9 & 0 & 0 \\ 0 & 50 & -100 & 50 & 0 & 0 & 0.9 & 0 \\ 0 & 0 & 50 & -50 & 0 & 0 & 0 & 0.9 \end{bmatrix}$$

$$\mathbf{B}(k) = \begin{bmatrix} 0 & 0 & 0 & 0 \\ 0 & 0 & 0 & 0 \\ 0 & 0 & 0 & 0 \\ 0 & 0 & 0 & 0 \\ 1 & 0 & 0 & 0 \\ 0 & 1 & 0 & 0 \\ 0 & 0 & 1 & 0 \\ 0 & 0 & 0 & 1 \end{bmatrix}$$

$$\mathbf{H}(k) = \begin{bmatrix} 1 & 0 & 0 & 0 & 0 & 0 & 0 & 0 \\ 0 & 1 & 0 & 0 & 0 & 0 & 0 & 0 \\ 0 & 0 & 1 & 0 & 0 & 0 & 0 & 0 \\ 0 & 0 & 0 & 1 & 0 & 0 & 0 & 0 \end{bmatrix}$$

$$\mathbf{R}(k) = \begin{bmatrix} 2 & 0 & 0 & 0 \\ 0 & 1.5 & 0 & 0 \\ 0 & 0 & 4 & 0 \\ 0 & 0 & 0 & 2.5 \end{bmatrix}.$$

## 5.4.3  Nodal Transformation Matrices

$$\mathbf{T}_0^l(k) = \begin{bmatrix} 4 & 3 & 0 & 0 & 0 & 0 & 0 & 0 \\ 2 & 0 & 0 & 0 & 0 & 0 & 0 & 0 \\ 0 & 6 & 0 & 0 & 0 & 0 & 0 & 0 \\ 0 & 0 & 0 & 0 & 4 & 2.5 & 0 & 0 \\ 0 & 0 & 0 & 0 & 0 & 7.5 & 0 & 0 \end{bmatrix}$$

$$= \mathcal{T}_0(k)\mathbf{T}_0(k)$$

$$= \begin{bmatrix} 1 & 1 & 0 & 0 \\ 0.5 & 0 & 0 & 0 \\ 0 & 2 & 0 & 0 \\ 0 & 0 & 1 & 1 \\ 0 & 0 & 0 & 3 \end{bmatrix} \begin{bmatrix} 4 & 0 & 0 & 0 & 0 & 0 & 0 & 0 \\ 0 & 3 & 0 & 0 & 0 & 0 & 0 & 0 \\ 0 & 0 & 0 & 0 & 4 & 0 & 0 & 0 \\ 0 & 0 & 0 & 0 & 0 & 2.5 & 0 & 0 \end{bmatrix}$$

$$
\mathbf{T}_1(k) =
\begin{bmatrix}
6.5 & 0 & 0 & 0 & 0 & 0 & 0 & 0 \\
0 & 3 & 0 & 0 & 0 & 0 & 0 & 0 \\
0 & 0 & 0 & 0 & 4 & 0 & 0 & 0 \\
0 & 0 & 0 & 0 & 0 & 5 & 0 & 0
\end{bmatrix}
$$

$$
\mathbf{T}_2(k) =
\begin{bmatrix}
0 & 0 & 11 & 0 & 0 & 0 & 0 & 0 \\
0 & 0 & 0 & 7 & 0 & 0 & 0 & 0 \\
0 & 0 & 0 & 0 & 0 & 0 & 4 & 0 \\
0 & 0 & 0 & 0 & 0 & 0 & 0 & 5.5
\end{bmatrix}
$$

$$
\mathbf{T}_3^l(k) =
\begin{bmatrix}
0 & 0 & 2 & 2.5 & 0 & 0 & 0 & 0 \\
0 & 1.5 & 0 & 0 & 0 & 0 & 0 & 0 \\
0 & 0 & 4 & 0 & 0 & 0 & 0 & 0 \\
0 & 0 & 0 & 0 & 0 & 0 & 4.5 & 6 \\
0 & 0 & 0 & 0 & 0 & 0 & 0 & 3
\end{bmatrix}
$$

$$
= \mathcal{T}_3(k)\mathbf{T}_3(k)
$$

$$
=
\begin{bmatrix}
0 & 0.5 & 0 & 0 \\
1 & 0 & 0 & 0 \\
0 & 1 & 0 & 0 \\
0 & 0 & 1 & 2 \\
0 & 0 & 0 & 1
\end{bmatrix}
\begin{bmatrix}
0 & 1.5 & 0 & 0 & 0 & 0 & 0 & 0 \\
0 & 0 & 4 & 0 & 0 & 0 & 0 & 0 \\
0 & 0 & 0 & 0 & 0 & 0 & 4.5 & 0 \\
0 & 0 & 0 & 0 & 0 & 0 & 0 & 3
\end{bmatrix} .
$$

### 5.4.4 Local Discrete Time Models

$$
\mathbf{F}_0(k) = \mathbf{T}_0(k)\mathbf{F}(k)\mathbf{T}_0^+(k)
$$

$$
=
\begin{bmatrix}
1 & 0 & 1 & 0 \\
0 & 1 & 0 & 1.2 \\
-50 & 66.7 & 0.9 & 0 \\
31.3 & -83.3 & 0 & 0.9
\end{bmatrix}
$$

Similarly,

$$
\mathbf{F}_1(k) =
\begin{bmatrix}
1 & 0 & 1.63 & 0 \\
0 & 1 & 0 & 0.6 \\
-30.8 & 66.7 & 0.9 & 0 \\
38.5 & -166.7 & 0 & 0.9
\end{bmatrix}
$$

$$\mathbf{F}_1(k) = \begin{bmatrix} 1 & 0 & 1 & 1 \\ 0 & 1 & 0 & 1.2 \\ -50 & 66.7 & 0.9 & 0 \\ 31.25 & -83.3 & 0 & 0.9 \end{bmatrix}$$

$$\mathbf{F}_2(k) = \begin{bmatrix} 1 & 0 & 2.75 & 0 \\ 0 & 1 & 0 & 1.27 \\ -36.4 & 28.6 & 0.9 & 0 \\ 25 & -39.3 & 0 & 0.9 \end{bmatrix}$$

$$\mathbf{F}_3(k) = \begin{bmatrix} 1 & 0 & 0 & 0 \\ 0 & 1 & 0.89 & 0 \\ 150 & -112.5 & 0.9 & 0 \\ 0 & 37.5 & 0 & 0.9 \end{bmatrix}$$

$$\mathbf{C}_0^l(k) = \begin{bmatrix} 1 & 1 & 0 & 0 & 0 \\ 0 & 1 & 0 & 0 & 0 \end{bmatrix}$$

$$\mathbf{C}_1(k) = \begin{bmatrix} 1 & 0 & 0 & 0 \end{bmatrix}$$

$$\mathbf{C}_2(k) = \begin{bmatrix} 1 & 0 & 0 & 0 \\ 0 & 1 & 0 & 0 \end{bmatrix}$$

$$\mathbf{C}_3^l(k) = \begin{bmatrix} 0 & 0 & 1 & 0 & 0 \end{bmatrix}$$

## 5.5   Summary

The stochastic control problem and its solution have been discussed for both linear and nonlinear systems. Using the separation principle for linear systems and assumed certainty equivalence for nonlinear systems, the decentralized estimation algorithms of Chapters 3 and 4 were extended to decentralized control systems. The algorithms from Chapter 3 were used to derive fully connected decentralized control algorithms. The drawbacks of these algorithms were eliminated by distributing control models while using decentralized estimation algorithms from Chapter 4. The resulting algorithm allows the development of general, scalable and flexible control systems. Its major novelty is the model defined, non-fully connected nature of the network. There is no propagation of information between unconnected nodes. Control nodes can be added or removed without design or algorithm change.

# Chapter 6

## Multisensor Applications: A Wheeled Mobile Robot

### 6.1   Introduction

This chapter describes the implementation of the theory developed in this book to a modular *wheeled mobile robot*, **WMR**. The main objective is to demonstrate the practical validity of the essential theory. The starting point is the construction of a kinematic model for a general WMR vehicle. This is done by using plane motion kinematics to derive forward and inverse kinematics for a generalized simple wheeled vehicle. This is then used to develop a modular decentralized kinematic model, which is combined with the control algorithm in Chapter 5 to provide decentralized WMR control. The mechanical structure of the WMR is discussed. The vehicle on which the estimation and control algorithms are tested is then described. The issue of which WMR modules need to communicate and the information they need to exchange is then considered. The Transputer architecture is used as the basis for hardware and software design as it supports the extensive communication and concurrency requirements characteristic of modular and decentralized systems. The modular software design is then discussed in detail. All the software is written in Parallel ANSI C and consists of two main parts: a configuration program and a nodal program which is loaded at each module. Pseudocode is used to demonstrate how the software achieves concurrency, modularity and local internodal communication. Examples of trajectories generated for the WMR are then outlined. The results of this implementation are described in Chapter 7.

## 6.2    Wheeled Mobile Robot (WMR) Modeling

The starting point in the modeling process is understanding the kinematics of the WMR. In general, kinematics refers to the branch of dynamics which treats motion without regard to forces which cause it. The motion of a WMR must be completely modeled at a kinematic level to enable it to perform a task or reach a goal. Several methods have been used to model the kinematics of WMRs [4], [5], [33], [39], [78], [79]. The modeling employed in this book is based on the work of Burke [33], and Alexander and Maddocks [4], [5]. The modeling principle is to ensure sufficient generality so that the technique is applicable to any mobile robot with simple wheels.

A WMR is modeled as a planar rigid robot body that moves over a horizontal reference plane on wheels that are connected to the body by axles. The role of the body of the WMR is to carry a moving coordinate system. An axle is supported by a single simple wheel, which is idealized as a disc without thickness, that lies in a vertical plane through the axle point. For a WMR the kinematic transform of interest relates the motion of the wheels and the WMR body. The motion of the WMR, an omni-directional vehicle, at a discrete time step $k$, is defined by the state vector

$$\mathbf{x}_b(k) = \begin{bmatrix} V_b(k) \\ \gamma_b(k) \\ \dot{\phi}_b(k) \end{bmatrix}, \tag{6.1}$$

where $V_b(k)$ is the WMR body drive velocity, $\gamma_b(k)$ is the body steer angle and $\dot{\phi}_b(k)$ is the rate of change of body orientation. The observed motion of a wheel $i$ at any specified position on the WMR at time $k$ is defined by the observation vector,

$$\mathbf{z}_i(k) = \begin{bmatrix} V_i(k) \\ \gamma_i(k) \end{bmatrix}, \tag{6.2}$$

where $V_i(k)$ is the wheel drive velocity and $\gamma_i(k)$ the wheel steer angle. The observed motion of all the wheels on a WMR at time $k$ is defined by the observation vector,

$$\mathbf{z}(k) = \begin{bmatrix} \mathbf{z}_0^T(k), \cdots, \mathbf{z}_{n-1}^T(k) \end{bmatrix}^T. \tag{6.3}$$

The transform which defines the motion of the WMR body, given the observed action of the wheels, is called the forward kinematics. It is used to compute an estimate of WMR body state vector $\hat{\mathbf{x}}_b(k \mid k)$ given sensed wheel parameters $\mathbf{z}_i(k)$. In general, forward kinematics can only be computed by a combination of observations from more than one wheel. The kinematic transform which defines the action of a general wheel $i$, given

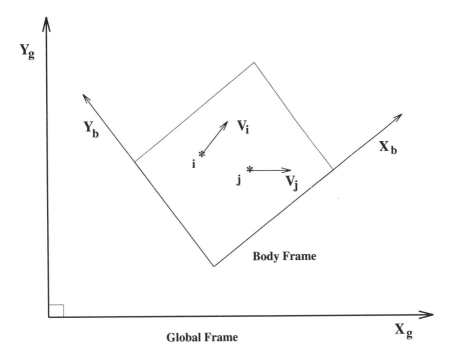

**FIGURE 6.1**
**A Rigid Body Under Plane Motion**

the motion of the WMR body, is called the inverse kinematics. The inverse kinematics of the WMR, for wheel $i$, is defined by a set of nonlinear equations of the form,

$$\mathbf{z}_i(k) = \mathbf{h}_i(\mathbf{x}_b(k)), \tag{6.4}$$

where $\mathbf{z}_i(k)$ is the motion of wheel $i$ at time $k$, and $\mathbf{x}_b(k)$ is the motion of the WMR body at time $k$, and $\mathbf{h}_i$ is the nonlinear inverse kinematic model. Unlike the forward kinematics, this transform usually provides a unique solution and can be solved in closed form.

## 6.2.1  Plane Motion Kinematics

The kinematics of any WMR is derived from basic theory of *planar* motion. This approach is due to Alexander and Maddocks [4], [5] and is developed in the work of Burke [33]. An inverse kinematics solution relates the motion of two wheels and then relates the action of the body to this motion. A laminar executes *plane motion* when all parts of the laminar move in parallel planes. There are two main types of plane motion, *rotation* and *translation*. Rotation about a fixed axis is the angular motion about the axis such that all parts of the laminar move in circular paths about the axis of rotation and all lines in the laminar rotate through the same angle at the same time. Translation is defined as any motion in which every line in the laminar remains parallel to its original position at all times. General planar motion may be a pure translation, a pure rotation, or a combination of both. The kinematics of a WMR has to obey two important constraints, the *rigid body* and *rolling* conditions.

**Rigid Body Condition**

A *rigid body* is defined as a system of particles for which the distance between the particles remains unchanged. Thus if each particle of a laminar is located by a position vector from a reference axis attached to and rotating with the body, there would be no change in the vector as measured from this axis. Consider a rigid body under plane motion as shown in Figure 6.1. A global coordinate system **G** is defined to be the stationary coordinate system with the z-axis orthogonal to the plane of travel. A body coordinate system **B** moves with the laminar and is also orthogonal to the plane of travel. Consider two points on the body, $i$ and $j$ with positions $p_i$ and $p_j$. If point $i$ moves with velocity $V_i$ and point $j$ moves with velocity $V_j$, then the *Rigid body condition* is stated as

$$(p_i - p_j).(V_i - V_j) = 0. \tag{6.5}$$

This means that points are fixed in the body frame and hence the distance between them remains constant.

**Rolling Condition**

The angular velocity at any point on the laminar is the same. At every instant the motion of the laminar coincides with either a pure translation or pure rotation about some point axis that is orthogonal to the laminar. This axis is called an *instantaneous center of rotation* (ICR). The movement or *rolling* about the ICR is specified if all points on the body are moving with the same angular velocity. If a point is at an angle $\phi$ in the global frame, then using the general points $i$ and $j$, the *Rolling condition* can be stated as

$$\frac{d\phi_i}{dt} = \frac{d\phi_j}{dt} = \frac{d\phi_b}{dt}, \tag{6.6}$$

where $\frac{d\phi_b}{dt}$ is the angular velocity of center of mass of the body. This means that all points on a rigid body move with the same angular speed.

Modeling the kinematics of a WMR can be achieved by deriving equations that satisfy the *rigid body* and *rolling* conditions for the WMR. The wheels of the WMR, connecting the body to the surface of travel are considered

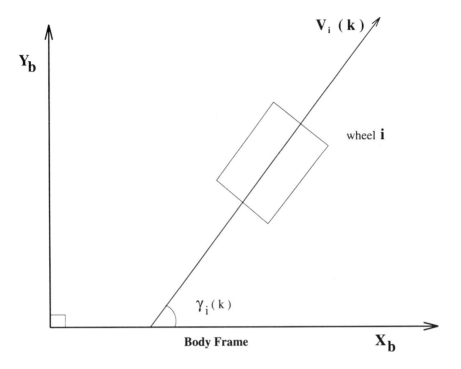

**FIGURE 6.2**
**A Simple Wheel in a Body Frame**

as points on the laminar. The wheels of the WMR move along the floor, which is accepted to be planar, hence the WMR body also moves in the same plane.

### 6.2.2   Decentralized Kinematics

**Two simple wheels**

A simple wheel is defined as a disc of radius $r$, without thickness, which lies in a vertical plane (about its center point) through the axle point. It can rotate in its vertical plane about its center point. It can be driven, steered, or both. Simple wheels are the most appropriate type of wheel for a domestic, industrial or office WMR application. If a WMR is considered as a rigid body and it moves with rolling motion on a plane surface, then the plane kinematics of rigid bodies may be used to relate the action of one wheel to another. Figure 6.2 shows one simple wheel in the WMR body frame. Its motion is characterized by two parameters, velocity $V_i(k)$ and

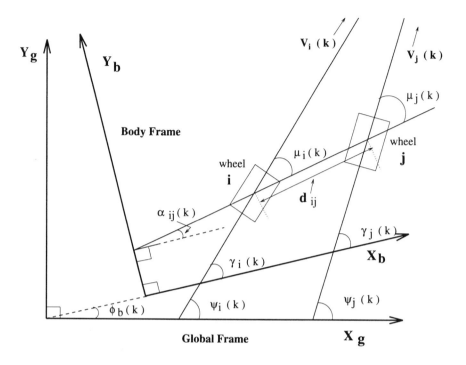

**FIGURE 6.3**
**Two Simple Wheels in a WMR**

steer angle $\gamma_i(k)$. Consider two such simple wheels, $i$ and $j$ in a WMR within a global frame, driving with velocities $V_i(k)$ and $V_j(k)$, as depicted in Figure 6.3. The wheels point at *physical* steer angles $\gamma_i(k)$ and $\gamma_j(k)$, respectively. The constant angle of the line connecting $i$ to $j$ is $\alpha_{ij}$ in the body frame. The magnitude of the distance between $i$ and $j$ is $d_{ij}$. The angle $\mu_i(k)$ is the *relative* steer angle of wheel $i$ to the direction $\alpha_{ij}$ and is given by

$$\mu_i(k) = \gamma_i(k) - \alpha_{ij}. \tag{6.7}$$

For a rigid body Equation 6.5 must be satisfied. This means that the component of $V_i(k)$ in the direction $\alpha_{ij}$ must be equal to the component of $V_j(k)$ in the same direction.

$$V_i(k) \cos \mu_i(k) - V_j(k) \cos \mu_j(k) = 0. \tag{6.8}$$

For the rolling condition to be satisfied, the normals of the velocity directions of every wheel must intersect at the ICR at any given moment. The angular velocity of any line in the body is equal to the angular velocity of

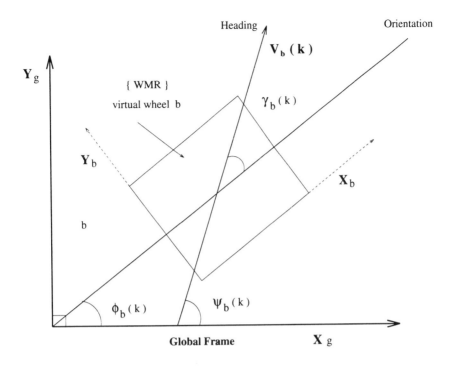

**FIGURE 6.4**
**The Concept of a Virtual Wheel**

the body. This is stated as follows:

$$\dot{\phi}_b = \frac{V_i(k)\sin\mu_i(k) - V_j(k)\sin\mu_j(k)}{d_{ij}}. \tag{6.9}$$

From Equations 6.8 and 6.9 there are four parameters, $V_i(k)$, $V_j(k)$, $\mu_i(k)$ and $\mu_j(k)$; three of which are independent. If three of these parameters are specified, then the motion of the WMR is completely defined. From Equation 6.9 it is evident that wheel parameters from at least two wheels are needed to estimate the motion of the body of the WMR. Equations 6.8 and 6.9 can then be used to relate the state $\mathbf{z}_i(k)$ to the state $\mathbf{z}_j(k)$, of any other wheel $j$. If the state of wheel $i$ is known and the WMR has a known angular velocity of $\dot{\phi}_b(k)$, then the state of any other wheel on the body can be found by a combination of Equations 6.8 and 6.9. The velocity of wheel $j$ in terms of $\mathbf{z}_i(k)$ is thus defined by

$$V_j(k) = V_i(k)\sqrt{\left[\cos^2\mu_i(k) + \left(\frac{\dot{\phi}_b d_{ij}}{V_i(k)} + \sin\mu_i(k)\right)^2\right]}. \tag{6.10}$$

The relative steer angle of wheel $j$ in terms of $\mathbf{z}_i(k)$ is defined as

$$\mu_j(k) = \arctan\left( \frac{\dot{\phi}_b(k)d_{ij}+V_i(k)\sin\mu_i(k)}{V_i(k)\cos\mu_i(k)} \right). \tag{6.11}$$

The physical steer angle $\gamma_j(k)$ is found by rearranging Equation 6.7 to give

$$\gamma_j(k) = \mu_j(k) + \alpha_{ij}. \tag{6.12}$$

**Virtual Wheel**

Kinematically, the body of the WMR can be represented by a *virtual wheel*. This is an imaginary wheel located at the point $b$ on the WMR body, as shown in Figures 6.4 and 6.5. If the WMR has a motion defined by the state vector $\mathbf{x}_b(k)$, then the virtual wheel has a corresponding motion $\mathbf{z}_b(k)$. The motion of any other wheel on the WMR can then be found from Equations 6.10 and 6.11. The velocity of the virtual wheel is set to be the desired velocity of the WMR body. From Figure 6.4, it can be seen that the physical steer angle of the virtual wheel $\gamma_b(k)$ is given by

$$\gamma_b(k) = \psi_b(k) - \phi_b(k). \tag{6.13}$$

The steer angle of the virtual wheel is the difference between the heading of the virtual wheel $\psi_b(k)$ and the orientation $\phi_b(k)$ of the WMR, from Equation 6.13. The angle $\gamma_b(k)$ is directly computed from the forward kinematics as a steer angle of another wheel.

**Modular Inverse kinematics**

The next step is to derive modular inverse kinematics for a general wheel $i$ given only knowledge of the position of that wheel with respect to the virtual wheel. Figure 6.5 illustrates the relation between the virtual wheel and a general wheel $i$. The inverse kinematic function $\mathbf{h}_i$ for any wheel $i$ on the WMR given in Equation 6.4 is solved by using Equations 6.10 and 6.11. The relative steer angle of the virtual wheel to the direction $\alpha_{bi}$ is given by

$$\mu_b(k) = \gamma_b(k) - \alpha_{bi}. \tag{6.14}$$

If the distance between point $b$ (the virtual wheel) and any wheel $i$ is $d_{bi}$, as depicted in Figure 6.5, then the velocity of wheel $i$ is given by

$$V_i(k) = V_b(k)\sqrt{\left[ \cos^2\mu_b(k) + \left( \frac{\dot{\phi}_b(k)d_{bi}}{V_b(k)} + \sin\mu_b(k) \right)^2 \right]}, \tag{6.15}$$

while the relative steer angle of wheel $i$ is obtained from

$$\mu_i(k) = \arctan\left( \frac{\dot{\phi}_b(k)d_{bi}+V_b(k)\sin\mu_b(k)}{V_b(k)\cos\mu_b(k)} \right). \tag{6.16}$$

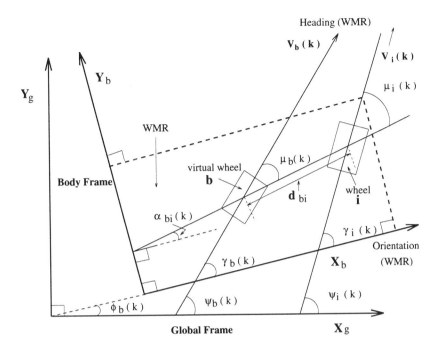

**FIGURE 6.5**
**The Virtual Wheel 'b' and a General Wheel 'i'**

The steer angle of wheel $i$ is defined by

$$\gamma_i(k) = \mu_i(k) + \alpha_{bi}. \tag{6.17}$$

## 6.3 Decentralized WMR Control

The inverse and the forward kinematics solutions complete the modular decentralized kinematic model of the WMR. This section describes how the modular decentralized kinematic model is then used in the scalable decentralized estimation and control algorithms described in Chapters 4 and 5 to produce a decentralized WMR controller. Since the vehicle model is nonlinear, algorithms based on the DDEIF are used. Specific WMR system models used in the implementation are presented.

### 6.3.1   General WMR System Models

The global (central) state vector $\mathbf{x}(k)$ for a WMR system with $n$ wheels is defined as a vector consisting of all the velocities and steer angles of the WMR physical wheels.

$$
\mathbf{x}(k) = \begin{bmatrix} V_0(k) \\ \gamma_0(k) \\ \cdot \\ \cdot \\ V_i(k) \\ \gamma_i(k) \\ \cdot \\ \cdot \\ V_{n-1}(k) \\ \gamma_{n-1}(k) \end{bmatrix} = \begin{bmatrix} \mathbf{z}_0(k) \\ \cdot \\ \cdot \\ \mathbf{z}_i(k) \\ \cdot \\ \cdot \\ \mathbf{z}_{n-1}(k) \end{bmatrix} . \tag{6.18}
$$

The global state transition function is defined from the forward kinematics solution as a stacked vector operator of *nonlinear* state transition functions, mapping the previous global state and control vectors to the current state vector, i.e.,

$$
\mathbf{f}(\cdot,\cdot,k) = \begin{bmatrix} \mathbf{f}_0 \\ \cdot \\ \cdot \\ \mathbf{f}_i \\ \cdot \\ \cdot \\ \mathbf{f}_{n-1} \end{bmatrix} , \tag{6.19}
$$

where $\mathbf{f}_i(\cdot,\cdot,k)$ is the state transition function defined by the forward kinematics Equations 6.10, 6.11 and 6.12. The global reference (demand) state vector is given by

$$
\mathbf{x}_r(k) = \begin{bmatrix} \mathbf{z}_{r_0}(k) \\ \cdot \\ \cdot \\ \mathbf{z}_{r_i}(k) \\ \cdot \\ \cdot \\ \mathbf{z}_{r_{n-1}}(k) \end{bmatrix} = \begin{bmatrix} \mathbf{h}_0(\mathbf{x}_{b_r}(k)) \\ \cdot \\ \cdot \\ \mathbf{h}_i(\mathbf{x}_{b_r}(k)) \\ \cdot \\ \cdot \\ \mathbf{h}_{n-1}(\mathbf{x}_{b_r}(k)) \end{bmatrix} , \tag{6.20}
$$

where the reference vehicle body state vector is

$$\mathbf{x}_{b_r}(k) = \begin{bmatrix} V_{b_r}(k) \\ \gamma_{b_r}(k) \\ \dot{\phi}_{b_r}(k) \end{bmatrix}. \tag{6.21}$$

The vector operator $\mathbf{h}_i(\cdot, k)$ represents the nonlinear inverse kinematic function given in Equations 6.15, 6.16 and 6.17.

### The DSU Transformation Matrix

Consider the general case where the estimation and control functions are carried out locally at the DSU $i$ (driven and steered unit $i$). The local state vector vector $\mathbf{x}_i(k)$ at DSU $i$ contains all the states required to estimate and control the velocity $V_i(k)$ and steer angle $\gamma_i(k)$. A sufficient representation of $\mathbf{x}_i(k)$ is given by

$$\mathbf{x}_i(k) = \begin{bmatrix} V_i(k) \\ \gamma_i(k) \\ \cdot \\ \cdot \\ \mathbf{t}_{ij} \begin{bmatrix} V_i(k) \\ \gamma_i(k) \end{bmatrix} \\ \cdot \\ \cdot \\ \mathbf{t}_{i(m-1)} \begin{bmatrix} V_{n-1}(k) \\ \gamma_{n-1}(k) \end{bmatrix} \end{bmatrix} = \begin{bmatrix} \mathbf{z}_i(k) \\ \cdot \\ \mathbf{t}_{ij}\mathbf{z}_j(k) \\ \cdot \\ \cdot \\ \mathbf{t}_{i(m-1)}\mathbf{z}_{m-1}(k) \end{bmatrix}, \tag{6.22}$$

where $m$ is less than or equal to $n$ and $\mathbf{t}_{ij}$ is an internodal transformation function defined by the forward kinematics in Equations 6.10, 6.11 and 6.12.

Applying the model distribution Equation 4.1, produces a general, stacked DSU transformation vector operator consisting of nonlinear transformation functions,

$$\mathbf{t}_i(\cdot, k) = \begin{bmatrix} 1 \\ \cdot \\ \cdot \\ \mathbf{t}_{ij} \\ \cdot \\ \cdot \\ \mathbf{t}_{i(m-1)} \end{bmatrix}, \tag{6.23}$$

where $\mathbf{t}_{ij}(\cdot, k)$ is a nonlinear transformation function relating wheel $i$ to wheel $j$. The function $\mathbf{t}_{ij}$ is a $\mathbf{0}$ function if observation information from DSU $j$ is not required at DSU $i$. In deciding which DSUs communicate, the overriding factor is that each DSU $i$ must be able to compute the WMR body state estimate $\hat{\mathbf{x}}_b(k \mid k)$. To do this, information from at least two wheels is required. Using the WMR nodal (DSU) transformation matrix, $\mathbf{t}_i(\cdot, k)$, the global WMR models are distributed using Equations 4.32, 4.33 and 5.21. The local reference vector is given by

$$
\mathbf{x}_{r_i}(k) =
\begin{bmatrix}
\mathbf{z}_{r_i}(k) \\
\cdot \\
\cdot \\
\mathbf{0}_i \\
\cdot \\
\cdot \\
\mathbf{0}_m
\end{bmatrix}
=
\begin{bmatrix}
\mathbf{h}_i(\mathbf{x}_{b_r}(k)) \\
\cdot \\
\cdot \\
\mathbf{0}_i \\
\cdot \\
\cdot \\
\mathbf{0}_m
\end{bmatrix} .
\tag{6.24}
$$

With local models and vectors for each DSU, the distributed and decentralized algorithm of Chapter 5 can then be used to provide scalable decentralized WMR control. Figure 6.8 shows the local control at a general driven and steered unit, DSU $i$.

### 6.3.2  Specific WMR Implementation Models

The WMR vehicle used has three driven and steered wheels, thus the global state vector is given by

$$
\mathbf{x}(k) =
\begin{bmatrix}
\mathbf{z}_0(k) \\
\mathbf{z}_1(k) \\
\mathbf{z}_2(k)
\end{bmatrix}
=
\begin{bmatrix}
V_0(k) \\
\gamma_0(k) \\
V_1(k) \\
\gamma_1(k) \\
V_2(k) \\
\gamma_2(k)
\end{bmatrix} .
\tag{6.25}
$$

The global nonlinear state transition matrix is defined from forward kinematics such that

$$
\mathbf{f}(\cdot, \cdot, k) =
\begin{bmatrix}
\mathbf{f}_0 \\
\mathbf{f}_1 \\
\mathbf{f}_2
\end{bmatrix} ,
\tag{6.26}
$$

and the process noise model is given by

$$
\mathbf{Q}(k) = \mathrm{E}\left[\Delta\mathbf{x}(k)\Delta\mathbf{x}^T(k)\right] = \nabla\mathbf{f}_x(k)\Sigma_k\nabla\mathbf{f}_x^T(k),
$$

where

$$
\mathbf{f}\left(\mathbf{x}(k-1), \mathbf{u}(k-1), (k)\right) =
$$

$$
\begin{bmatrix}
V_0(k-1)+ \\
V_1(k-1)\sqrt{\left[\cos^2\left[\gamma_1(k-1)-\alpha_{01}\right]+\left(\frac{\dot{\phi}_b d_{01}}{V_1(k-1)}+\sin\left[\gamma_1(k-1)-\alpha_{01}\right]\right)^2\right]}+ \\
V_2(k-1)\sqrt{\left[\cos^2\left[\gamma_2(k-1)-\alpha_{02}\right]+\left(\frac{\dot{\phi}_b d_{01}}{V_2(k-1)}+\sin\left[\gamma_2(k-1)-\alpha_{02}\right]\right)^2\right]} \\[2ex]
\gamma_0(k-1)+\arctan\left(\frac{\dot{\phi}_b(k-1)d_{01}+V_1(k-1)\sin\left[\gamma_1(k-1)-\alpha_{01}\right]}{V_1(k)\cos\left[\gamma_1(k-1)-\alpha_{01}\right]}\right)+\alpha_{01}+ \\
\arctan\left(\frac{\dot{\phi}_b(k)d_{20}+V_2(k)\sin\left[\gamma_2(k)-\alpha_{20}\right]}{V_2(k)\cos\left[\gamma_2(k)-\alpha_{20}\right]}\right)+\alpha_{20} \\[2ex]
V_1(k-1)+ \\
V_0(k-1)\sqrt{\left[\cos^2\left[\gamma_0(k-1)-\alpha_{01}\right]+\left(\frac{\dot{\phi}_b d_{01}}{V_0(k-1)}+\sin\left[\gamma_0(k-1)-\alpha_{01}\right]\right)^2\right]}+ \\
V_2(k-1)\sqrt{\left[\cos^2\left[\gamma_2(k-1)-\alpha_{21}\right]+\left(\frac{\dot{\phi}_b d_{01}}{V_2(k-1)}+\sin\left[\gamma_2(k-1)-\alpha_{21}\right]\right)^2\right]} \\[2ex]
\gamma_1(k-1)+\arctan\left(\frac{\dot{\phi}_b(k-1)d_{01}+V_0(k-1)\sin\left[\gamma_0(k-1)-\alpha_{01}\right]}{V_0(k)\cos\left[\gamma_0(k-1)-\alpha_{01}\right]}\right)+\alpha_{01}+ \\
\arctan\left(\frac{\dot{\phi}_b(k)d_{21}+V_2(k)\sin\left[\gamma_2(k)-\alpha_{21}\right]}{V_2(k)\cos\left[\gamma_2(k)-\alpha_{21}\right]}\right)+\alpha_{01} \\[2ex]
V_2(k-1)+ \\
V_0(k-1)\sqrt{\left[\cos^2\left[\gamma_0(k-1)-\alpha_{02}\right]+\left(\frac{\dot{\phi}_b d_{02}}{V_0(k-1)}+\sin\left[\gamma_0(k-1)-\alpha_{02}\right]\right)^2\right]}+ \\
V_1(k-1)\sqrt{\left[\cos^2\left[\gamma_1(k-1)-\alpha_{12}\right]+\left(\frac{\dot{\phi}_b d_{12}}{V_1(k-1)}+\sin\left[\gamma_1(k-1)-\alpha_{12}\right]\right)^2\right]} \\[2ex]
\gamma_2(k-1)+\arctan\left(\frac{\dot{\phi}_b(k-1)d_{02}+V_0(k-1)\sin\left[\gamma_0(k-1)-\alpha_{02}\right]}{V_0(k)\cos\left[\gamma_0(k-1)-\alpha_{02}\right]}\right)+\alpha_{02}+ \\
\arctan\left(\frac{\dot{\phi}_b(k)d_{12}+V_1(k)\sin\left[\gamma_1(k)-\alpha_{12}\right]}{V_1(k)\cos\left[\gamma_1(k)-\alpha_{12}\right]}\right)+\alpha_{12}
\end{bmatrix}
$$

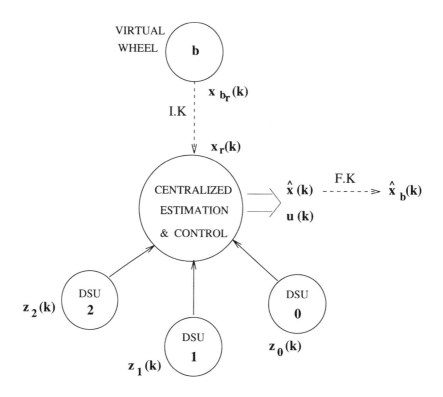

**FIGURE 6.6**
**WMR Centralized Control**

**Local DSU Models and Vectors**

The global models and vectors of the WMR are summarized in a centralized control configuration shown in Figure 6.6. Each DSU acts as a local observer with no estimation or control functions. The system is then distributed into a decentralized control system with local models. The local states are computed from the DSU transformation matrices.

$$\mathbf{x}_0(k) = \mathbf{t}_0 \begin{bmatrix} \mathbf{z}_0(k) \\ \mathbf{z}_1(k) \\ \mathbf{z}_2(k) \end{bmatrix}$$

$$= \begin{bmatrix} 1 \\ \mathbf{t}_{01} \\ \mathbf{t}_{02} \end{bmatrix} \begin{bmatrix} \mathbf{z}_0(k) \\ \mathbf{z}_1(k) \\ \mathbf{z}_2(k) \end{bmatrix}$$

$$= \begin{bmatrix} V_0(k) \\ \gamma_0(k) \\[2mm] V_1(k)\sqrt{\left[\cos^2\left[\gamma_1(k)-\alpha_{10}\right] + \left(\frac{\dot{\phi}_b d_{10}}{V_0(k)} + \sin\left[\gamma_1(k)-\alpha_{10}\right]\right)^2\right]} \\[2mm] \arctan\left(\frac{\dot{\phi}_b(k)d_{10}+V_1(k)\sin\left[\gamma_1(k)-\alpha_{10}\right]}{V_1(k)\cos\left[\gamma_1(k)-\alpha_{10}\right]}\right) + \alpha_{10}. \\[2mm] V_2(k)\sqrt{\left[\cos^2\left[\gamma_2(k)-\alpha_{20}\right] + \left(\frac{\dot{\phi}_b d_{20}}{V_2(k)} + \sin\left[\gamma_2(k)-\alpha_{20}\right]\right)^2\right]} \\[2mm] \arctan\left(\frac{\dot{\phi}_b(k)d_{20}+V_2(k)\sin\left[\gamma_2(k)-\alpha_{20}\right]}{V_2(k)\cos\left[\gamma_2(k)-\alpha_{20}\right]}\right) + \alpha_{20}. \end{bmatrix}.$$

Similarly, the local vectors of DSUs 1 and 2 are obtained.

$$\mathbf{x}_1(k) = \begin{bmatrix} V_0(k)\sqrt{\left[\cos^2\left[\gamma_0(k)-\alpha_{01}\right] + \left(\frac{\dot{\phi}_b d_{01}}{V_0(k)} + \sin\left[\gamma_0(k)-\alpha_{01}\right]\right)^2\right]} \\[2mm] \arctan\left(\frac{\dot{\phi}_b(k)d_{01}+V_0(k)\sin\left[\gamma_0(k)-\alpha_{01}\right]}{V_0(k)\cos\left[\gamma_0(k)-\alpha_{01}\right]}\right) + \alpha_{01}. \\[2mm] V_1(k) \\ \gamma_1(k) \\[2mm] V_2(k)\sqrt{\left[\cos^2\left[\gamma_2(k)-\alpha_{21}\right] + \left(\frac{\dot{\phi}_b d_{21}}{V_2(k)} + \sin\left[\gamma_2(k)-\alpha_{21}\right]\right)^2\right]} \\[2mm] \arctan\left(\frac{\dot{\phi}_b(k)d_{21}+V_2(k)\sin\left[\gamma_2(k)-\alpha_{21}\right]}{V_2(k)\cos\left[\gamma_2(k)-\alpha_{20}\right]}\right) + \alpha_{21}. \end{bmatrix}$$

$$\mathbf{x}_2(k) = \begin{bmatrix} V_0(k)\sqrt{\left[\cos^2\left[\gamma_0(k)-\alpha_{02}\right] + \left(\frac{\dot{\phi}_b d_{02}}{V_0(k)} + \sin\left[\gamma_0(k)-\alpha_{02}\right]\right)^2\right]} \\[2mm] \arctan\left(\frac{\dot{\phi}_b(k)d_{02}+V_0(k)\sin\left[\gamma_0(k)-\alpha_{02}\right]}{V_0(k)\cos\left[\gamma_0(k)-\alpha_{02}\right]}\right) + \alpha_{02} \\[2mm] V_1(k)\sqrt{\left[\cos^2\left[\gamma_1(k)-\alpha_{10}\right] + \left(\frac{\dot{\phi}_b d_{12}}{V_0(k)} + \sin\left[\gamma_1(k)-\alpha_{12}\right]\right)^2\right]} \\[2mm] \arctan\left(\frac{\dot{\phi}_b(k)d_{12}+V_1(k)\sin\left[\gamma_1(k)-\alpha_{12}\right]}{V_1(k)\cos\left[\gamma_1(k)-\alpha_{10}\right]}\right) + \alpha_{10}. \\[2mm] V_2(k) \\ \gamma_2(k) \end{bmatrix}.$$

As the network is fully connected, the local state transition matrices are of the same form for the three DSUs and the global system. The observation models are obtained from

$$\mathbf{H}_0(k) = \mathbf{C}_0(k) = [1 \quad 0 \quad 0] \tag{6.27}$$
$$\mathbf{H}_1(k) = \mathbf{C}_1(k) = [0 \quad 1 \quad 0] \tag{6.28}$$
$$\mathbf{H}_2(k) = \mathbf{C}_2(k) = [0 \quad 0 \quad 1]. \tag{6.29}$$

The local reference state vector for DSU 0 is derived from

$$\mathbf{x}_{r_0}(k) = \begin{bmatrix} \mathbf{h}_0(\mathbf{x}_{b_r}(k)) \\ 0 \\ 0 \end{bmatrix}$$

$$= \begin{bmatrix} \left[ V_{b_r}(k)\sqrt{\left[ \cos^2\left[\gamma_{b_r}(k) - \alpha_{b0}\right] + \left( \frac{\dot{\phi}_b(k)d_{b0}}{V_{b_r}(k)} + \sin\left[\gamma_{b_r}(k) - \alpha_{b0}\right] \right)^2 \right]} \right] \\ \arctan\left( \frac{\dot{\phi}_{b_r}(k)d_{b0} + V_{b_r}(k)\sin\left[\gamma_{b_r}(k) - \alpha_{b0}\right]}{V_{b_r}(k)\cos\left[\gamma_{b_r}(k) - \alpha_{b0}\right]} \right) + \alpha_{b0} \\ 0 \\ 0 \\ 0 \\ 0 \end{bmatrix}.$$

The reference state vectors of DSU 1 and 2 are found in a similar way.

$$\mathbf{x}_{r_1}(k)$$

$$= \begin{bmatrix} 0 \\ 0 \\ \left[ V_{b_r}(k)\sqrt{\left[ \cos^2\left[\gamma_{b_r}(k) - \alpha_{b1}\right] + \left( \frac{\dot{\phi}_{b_r}(k)d_{bi}}{V_{b_r}(k)} + \sin\left[\gamma_{b_r}(k) - \alpha_{b1}\right] \right)^2 \right]} \right] \\ \arctan\left( \frac{\dot{\phi}_{b_r}(k)d_{b1} + V_{b_r}(k)\sin\left[\gamma_{b_r}(k) - \alpha_{b1}\right]}{V_{b_r}(k)\cos\left[\gamma_{b_r}(k) - \alpha_{b1}\right]} \right) + \alpha_{b1} \\ 0 \\ 0 \end{bmatrix}$$

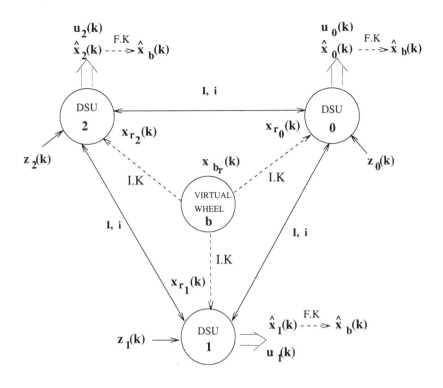

**FIGURE 6.7**
**WMR Decentralized Control**

$$\mathbf{x}_{r_2}(k)$$

$$= \begin{bmatrix} 0 \\ 0 \\ 0 \\ 0 \\ Vb_r(k)\sqrt{\left[\cos^2\left[\gamma_{b_r}(k)-\alpha_{b2}\right] + \left(\dfrac{\dot{\phi}_{b_r}(k)d_{bi}}{V_{b_r}(k)} + \sin\left[\gamma_{b_r}(k)-\alpha_{b2}\right]\right)^2\right]} \\ \arctan\left(\dfrac{\dot{\phi}_{b_r}(k)d_{b2}+V_{b_r}(k)\sin\left[\gamma_b(k)-\alpha_{b2}\right]}{V_{b_r}(k)\cos\left[\gamma_{b_r}(k)-\alpha_{b2}\right]}\right) + \alpha_{b2} \end{bmatrix}$$

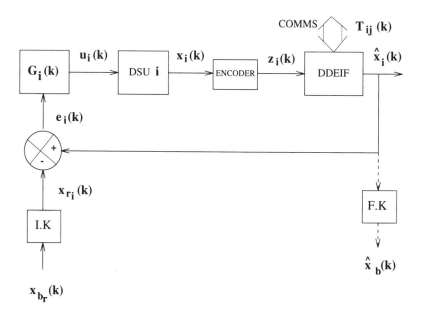

**FIGURE 6.8**
**WMR Decentralized Control**

### 6.3.3   Driven and Steered Unit (DSU) Control

Using the three DSU models described above, decentralized WMR control was implemented. Figure 6.7 summarizes the entire decentralized WMR control system. The terms I.K and F.K refer to the inverse and forward kinematic functions, respectively. Figure 6.6 shows the equivalent centralized control system which uses global vectors, models and a central processor. In the decentralized control system, each DSU $i$ estimates and controls a local state vector $\mathbf{x}_i(k)$ and not the vehicle body state vector $\mathbf{x}_b(k)$ or the global state vector $\mathbf{x}(k)$. In this way, the WMR control system is distinguished from that implemented in [33], where the body state vector is controlled. The closed loop control for each DSU is given in Figure 6.8. The control gain is applied to the difference between the local estimate and local reference state vectors. The local reference state vector is obtained from the WMR body reference (demand) state vector by inverse kinematics. From its local state vector estimate $\hat{\mathbf{x}}_i(k \mid k)$, each DSU can generate an estimate of the state vector of WMR body $\hat{\mathbf{x}}_b(k \mid k)$ using the forward kinematics Equations 6.10, 6.11 and 6.12.

### 6.3.4   Application of Internodal Transformation

The internodal transformation theory derived in Chapter 4 is used to formalize the design of a fully decentralized and modular WMR. This is done by providing a systematic way of deciding which active units need to communicate and what information they have to exchange. This is a very important consideration given hardware constraints, in particular the limitation of four links per Transputer (T805) (the local processor). In the WMR model described above, the internodal transformation from one DSU to the other is defined by the elements of the DSU transformation vector (operator). For example, the transformation vector at DSU 0 is given by

$$\mathbf{t}_0 = \begin{bmatrix} 1 \\ t_{01} \\ t_{02} \end{bmatrix}. \tag{6.30}$$

In Figure 6.7, the full connection between the three DSU is sufficient but not necessary to achieve decentralized WMR control. Only information from two wheels is strictly required at each DSU. Hence, the communication link between DSU 0 and 2 can be removed, reducing the size of the local state vectors at each of the two DSUs by two elements. For example, the state vector at DSU 0 is then computed as follows:

$$
\begin{aligned}
\mathbf{x}_0(k) &= \mathbf{t}_0 \begin{bmatrix} \mathbf{z}_0(k) \\ \mathbf{z}_1(k) \\ \mathbf{z}_2(k) \end{bmatrix} \\
&= \begin{bmatrix} 1 \\ t_{01} \end{bmatrix} \begin{bmatrix} \mathbf{z}_0(k) \\ \mathbf{z}_1(k) \\ \mathbf{z}_2(k) \end{bmatrix} \\
&= \begin{bmatrix} V_0(k) \\ \gamma_0(k) \\ V_1(k)\sqrt{\left[ \cos^2\left[\gamma_1(k) - \alpha_{10}\right] + \left( \frac{\dot{\phi}_b d_{10}}{V_0(k)} + \sin\left[\gamma_1(k) - \alpha_{10}\right] \right)^2 \right]} \\ \arctan\left( \frac{\dot{\phi}_b(k) d_{10} + V_1(k)\sin\left[\gamma_1(k) - \alpha_{10}\right]}{V_1(k)\cos\left[\gamma_1(k) - \alpha_{10}\right]} \right) + \alpha_{10} \end{bmatrix}.
\end{aligned}
$$

The results obtained with the non-fully connected WMR configuration are exactly the same as those obtained with the fully connected one, as discussed in Chapter 7.

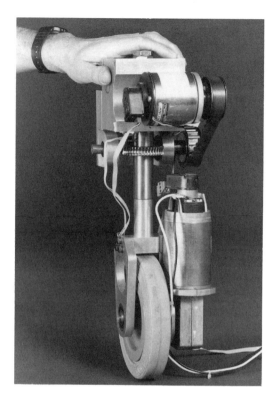

**FIGURE 6.9**
**A Driven and Steered Unit (DSU)**

## 6.4   Hardware Design and Construction

A modular vehicle has the same function as any conventional robot except that it is constructed from a small number of standard modular units.
These modular units are associated with the main vehicle components:
drive systems, sensing elements, actuators, power and communication. The
components are then assembled to effect the intended application and programmed accordingly. The main motivating benefits of modular robotics
are reduced prototyping and experimental costs, design flexibility and system reliability.

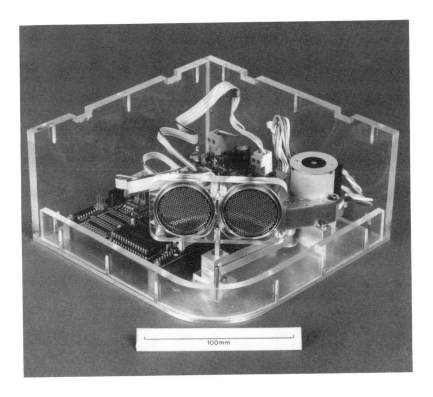

**FIGURE 6.10**
**A Differential Sonar Sensor**

### 6.4.1  WMR Modules

The WMR hardware described here is completely modular. Its main units, which will be briefly described, are

- A driven and steered unit

- A differential sonar sensor

- Chassis, battery and power units

- Communication units.

**A Driven and Steered Unit**

The motive power of the WMR is provided by a driven and steered unit (DSU), in which both steering and drive wheels are actuated. The steer actuator is mounted above the drive actuator. An encoder measures wheel speed on the drive motor and steer angle. An HCTL-1100 motor controller

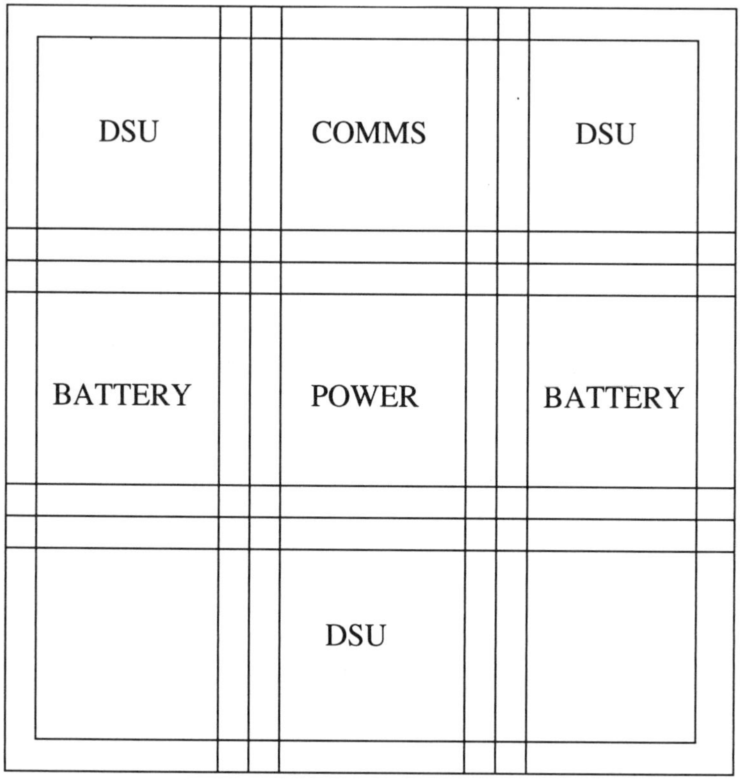

**FIGURE 6.11**
**Nine Connectable, Modular Chassis Frames**

board with a T805 Transputer is attached to the steering block. The DSU is shown in Figure 6.9.

**The Differential Sonar Sensor**

The WMR requires sensors for purposes of navigation and obstacle avoidance. A tracking differential sonar system based on the principle of region of constant depth detector differential time (RCDD) is used. Figure 6.10 shows such a tracker. It consists a pair of Polaroid electrostatic transducers mounted on a panning frame driven by a stepper motor. Two transducers are used as receivers for a single sonar pulse providing differential time of flight data. This provides real-time target bearing information needed for feature tracking. The RCDD is used to give positional updates of the WMR position.

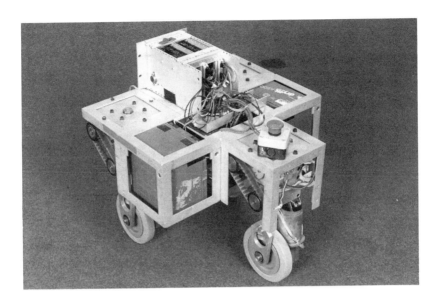

**FIGURE 6.12**
**A Modular WMR Composed of Three DSUs**

### Chassis, Battery and Power Units

A reconfigurable chassis is designed to hold different modular components. Its essential characteristic is adaptability to different WMR configurations, while being light and strong enough to hold all vehicle components. Figure 6.11 shows nine connectable modular chassis frames and possible WMR units they could hold. Rechargable gel lead acid batteries are used to power a 36v drive circuit and a 12v logic circuit. Different vehicle units have different power requirements and a power loom is employed to distribute power and to recharge the batteries on-board.

### Communication

There is extensive communication and concurrency between the different units of the WMR. Communication is based on Transputer communication links. The T805 Transputer has four physical communication links. Virtual links can also be employed. For sending commands to and logging data from the WMR, an RS422 link between the WMR an off-board PC is used.

## 6.4.2 A Complete Modular Vehicle

### A WMR Composed of Three DSUs

Six modular wheeled mobile robots were built as part of the OxNav

**FIGURE 6.13**
**Three DSU WMR: An Underside View**

project and two of these were used to test the theories developed in this book. The first vehicle consists of three DSUs, three battery units, power unit, communication units and sensors. Each DSU is as described above and constitutes the main active unit of the WMR. The DSUs communicate by use of Transputer architecture. There is no fixed axle, all three wheels are driven and steered. Thus the WMR is omni-directional, which means it can move in any direction irrespective of its posture. Figure 6.12 shows this WMR and a view of its underside is displayed in Figure 6.13.

**A WMR with Mounted RCDD Trackers**

In the second vehicle, four units of RCDD sonar trackers are mounted on top of the WMR. They are placed at the four corners of the top of the vehicle as shown in Figure 6.14. Each differential sonar sensor rotates and tracks environment features. By tracking the range and bearing to these features over time, two or more sonars can provide continuous updates of vehicle position. Since the vehicle is omni-directional, it can achieve any planar motion irrespective of its posture. The processors on the vehicle include three Transputers for the DSUs, three Transputers to run the sensor software and one Transputer to run the server process.

**WMR Design Scalability and Flexibility**

A vehicle was constructed to demonstrate the scalability, flexibility and

**FIGURE 6.14**
**A Complete WMR with Mounted RCDD Trackers**

adaptability of the modular system. It consists of eighteen modular chassis frames, six wheels and four DSUs as shown in Figure 6.15. Effectively, this vehicle is obtained by coupling a second WMR to the WMR which has mounted RCDD sensors, with both WMRs using the same software on component units. The processors on the scaled vehicle include four Transputers for the four DSUs, one Transputer to run software and two more to run server processes. There is a fixed axle and hence this WMR has 2DOF.

### 6.4.3   Transputer Architecture

Distributed hardware is required to implement the WMR. Decentralized fusion and control systems involve extensive communication and concurrency. Their hardware requirements include on-board processing power and memory, reliable communication, and efficient parallel processing. Transputer architecture meets these requirements. A Transputer is a single-chip VLSI device with its own processor, memory and communication links. It

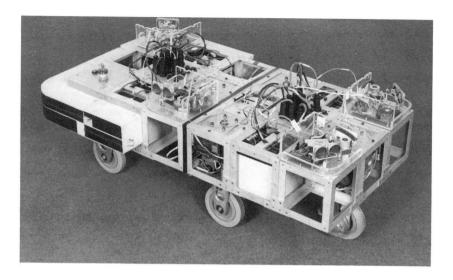

**FIGURE 6.15**
**Two Coupled WMRs: Illustration of Design Scalability**

may be connected to other Transputers in a network. It is a device designed for the implementation of multiprocessor systems by providing high speed, fast inter-processor communication and explicit support for multiple processes. The Transputer is thus an ideal device for decentralized architectures. It is designed to carry out external communication with other Transputers, combining a powerful processor with high level communication procedures. External communication takes place along the built-in serial links (with handshake) and is easily handled by high level programming languages such as Occam-2 or Parallel *ANSI C*.

The main features of a general Transputer are displayed in Figure 6.16. They include a high-speed integer processor, on-chip fast static memory, communication links and programmable external memory interface. A Transputer tram network consisting of TRAM2/T805 Transputers, each with 128Kb on-board memory running at either 10 or 20MHz, is used to carry out simulations of the algorithms developed in this book. The sensor input/output is handled by two C011 links to parallel interface chips. They each provide 8 bit input and output with two handshake bits. The Transputer network takes the form of trams mounted on a Personal Computer (PC). Figure 6.17 shows such a tram consisting of four fully connected Transputers. This network manifests complete asynchronous concurrency and communication.

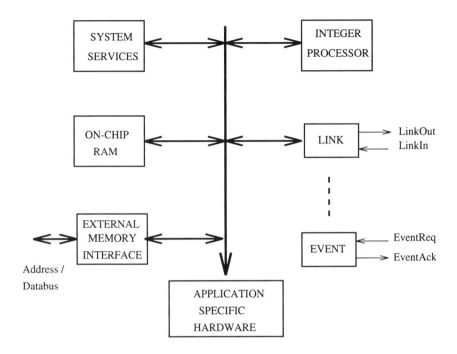

**FIGURE 6.16**
**A General Transputer**

## 6.5 Software Development

Software design is based on Transputer hardware and is developed in a modular and generalized fashion such that the same software is used on each DSU module. It was developed in such a way that software is interchangeable between simulations and the physical vehicle system. Although programming could have been done in either Occam-2 or Parallel ANSI C, the latter was chosen because it provides a good trade-off between communication capabilities and complex algorithm support. This language is essentially ANSI C with parallel libraries. The *Inmos* ANSI C Transputer toolset used supports concurrency and communication quite effectively while ensuring compatibility with existing software packages.

Structurally the software developed consists of two types of programs, the *configuration program* and the *nodal program*. The nodal program is a generic program running at each module (node) which is the same for all modules. It is modeled as a communicating control process, $CCP$.

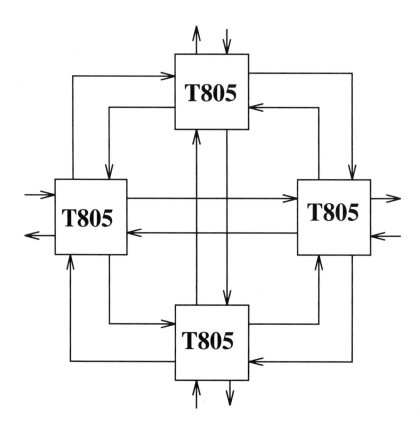

**FIGURE 6.17**
**A Four Transputer Tram Network**

The configuration file runs all the nodal programs in parallel and is modeled as an entire decentralized control system, $DCONT$ (fully connected) or $DDCONT$ (non-fully connected). Figure 6.18 displays an overview of the software modules, their internal processes and both external and internal communication requirements.

## 6.5.1   Nodal Program (Communicating Control Process)

This program carries out the nodal functions: estimation, communication and control. Each node has a generic communicating control process ($CCP$) as depicted in Figure 6.19. Locally, $CCP$ interacts with the local observer and system model. It inputs a local observation vector, processes information, communicates with other nodal $CCPs$, generates a control

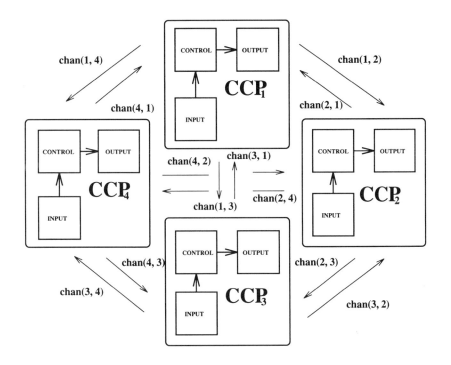

**FIGURE 6.18**
**Decentralized Software Development**

vector and feeds it back into the system. The taking of a local observation marks the beginning of an iteration, while the feedback of a control vector marks its end. Figure 6.19 shows the internal structure of the $CCP$ and its external channels. It contains three parallel communicating processes, $CONTROL$, $OUTPUT$ and $INPUT$. It has two internal channels $to.output$ and $from.input$.

The process $CONTROL$ communicates the local information state contribution $\mathbf{i}_i(k)$ and its associated local information matrix $\mathbf{I}_i(k)$ over channel $to.output$ to $OUTPUT$. The channel $from.input$ is a *bus-channel* which communicates $(N-1)$ information contributions and $(N-1)$ associated information matrices from $INPUT$ to $CONTROL$. Externally, $CCP$ communicates *exactly once* with each *other* node. It does not communicate with itself. Hence it has $(N-1)$ *sending* external channels through $OUTPUT$ and $(N-1)$ *receiving* external channels through $INPUT$. These channels are labeled as shown in Figure 6.19 such that $chan_{i,j}$ is a channel from $node_i$ to $node_j$. The software implementation of the $INPUT$ process is illustrated as follows:

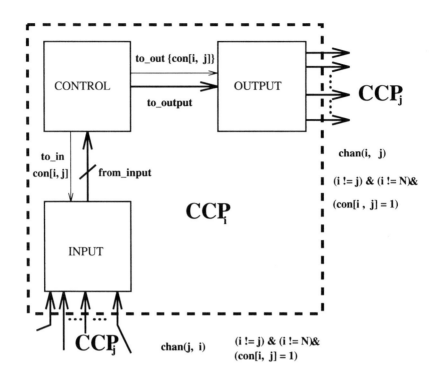

**FIGURE 6.19**
**Communicating Control Process**

```
                /* Declarations */
process: in, in_p
channels: from_in, from_input
variables: connect, info, INFO
                /* initialize variables */
info=0; INFO=0; connect=0;
/* If Nodes are connected, input information in parallel */
    ChanInQMat(from_in, connect);
    for (i = 0, p = 0, j = 0; i < Num_Nodes; i++) {
        if (i != Node_id) {
            if (e(connect, i, 0) != 0) {
            in_p[j] = ProcAlloc(in, 0, 4, chan_in[p],
                                info[i], INFO[i], i);
                    j++;}
                p++;}}
```

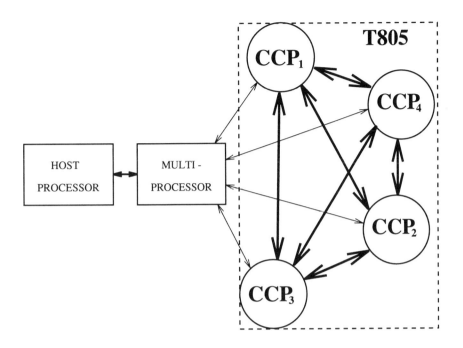

**FIGURE 6.20**
**A Single Transputer Configuration**

```
/* Send information to CONTROL */
    for (j; j < Num_Nodes; ++j) {
        in_p[j] = NULL;}
        while (TRUE) {
                ProcParList(in_p);
                for (i = 0; i < Num_Nodes; ++i) {
                    if (e(connect, i, 0) != 0) {
                        ChanOutMat(from_input, info[i]);
                        ChanOutMat(from_input, INFO[i]);}}}
```

**Internodal Transformation**

With model distribution communication is minimized both in terms of number of communication links and size of message. This is because communication then depends on connectedness, which in turn depends on internodal transformations. Consequently, only relevant information is exchanged. Two nodes $i$ and $j$ will communicate if, and only if, they have an overlapping information space; that is, at least one of the internodal transformation matrix $\mathbf{T}_{ij}(k)$ and $\mathbf{T}_{ji}(k)$ is not a null matrix. This condition

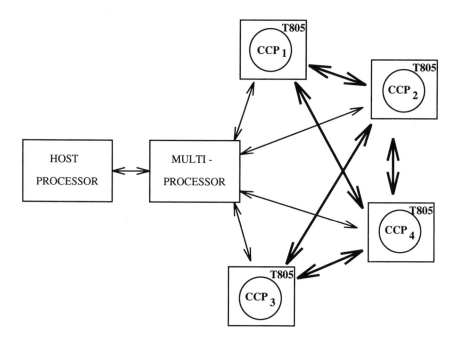

**FIGURE 6.21**
**A Four Transputer Configuration**

is illustrated in Figure 6.19 and lays the basis for software implementation of scalable decentralized control, $DDCONT$. The pseudocode below illustrates how internodal transformation is implemented in software.

```
/*Compute internodal transformation*/
  for (k = 0; k < Num_Nodes; k++) {
      if (k != Node_id) {
  qmat_mpy(T[k], temp4, Ti[k], T[k]->r, iXDIM, XDIM);
  /*I it is not null connect the nodes*/
      for (i = 0; i < T[k]->r; ++i){
          for (j = 0; j < XDIM; ++j) {
              if (e(Ti[k], i, j) == 1.0) {
              e(connect, k, 0) = 1.0;
              }}}}}
      /* Send result to INPUT and OUTPUT */
      ChanOutMat(to_out, connect);
      ChanOutMat(from_in, connect);
```

## 6.5.2   Configuration Program (Decentralized Control)

The configuration program (file) loads the nodal programs, $CCPs$ onto the Transputer network. The configuration file contains information about the number of communicating nodal programs and their interactions and establishes the network of communicating control processes as either fully connected or non-fully connected. It connects the *host* processor to the network and maps programs to processors. To distinguish between nodes, the configuration file uses a global vector with the same number of elements as the number of nodes, such that each element corresponds to a different node. The configuration program is implemented as shown below:

```
/*number of nodes in system*/
val Num_Nodes 4;
/* Host link connection */
connect PROC[0].link[BootLink] to host;
/* Connect processes using channels */
process (interface (input HostInput,
                    output HostOutput,
                    input from_control10,
                    output to_control01,
                    input from_control20,
                    output to_control02,
                    input from_control30,
                    output to_control03, int id)) control0;
/*Set Node identity */
control0(id=0);
control1(id=1);
control2(id=2);
control3(id=3)
/* Mapping programs to processors */
use "control.c9x" for control0;
use "control.c9x" for control1;
use "control.c9x" for control2;
use "control.c9x" for control3;
use "hostmult.c9x" for HostMult;
```

Figure 6.18 illustrates a fully connected network of four nodal control processes. It shows the four $CCPs$, their internal processes, internal communications and external communications. For simulations both nodal and configuration programs can be run on a single Transputer. Figure 6.20 shows a fully connected network of four communicating control nodes running on a single T805 Transputer, through a multiprocessor. Alternatively, each node can be run on its own Transputer. Figure 6.21 shows a non-fully connected network of four control nodes.

Nodes can be added or removed without changing the nodal program. Only the configuration file is changed to couple or decouple the node of interest.

## 6.6  On-Vehicle Software

Vehicle software used can be placed into two general categories, *user interface* software and *on-board* software. User interface software is used to specify WMR paths, set control parameters, select maps of the environment, process constants and WMR motion parameters. It also enables the appropriate on-board software to be selected, executed and terminated.

On-vehicle software describes software running on active units, sonar sensors, DSUs and communication units. This software is developed in a modular fashion with a vehicle configuration file and a nodal program running at each vehicle module. It is run on a Transputer network where each active drive unit, sensor node and communication unit contains a Transputer. The vehicle configuration program arranges the modular software so that it runs on a particular WMR. This is done by specifying the number of modules, their identities, inter-module communication, type of Transputer and positions of active units.

### 6.6.1  Nodal Software

The nodal software running at each active unit is identical for all units. It is modeled by a communicating control process, $CCP$, containing three parallel processes: $INPUT$, $OUTPUT$ and $CONTROL$. Figure 6.22 shows the structure of process $CONTROL$ with specific reference to WMR software. It consists of the following main processes running in parallel at set frequencies:

- $START$. This process coordinates the initialization of parallel processes. The server process uses this to synchronize the start of all the decentralized modular control programs.

- $PATHCONT$. The purpose of this process is to plan and track a path for the vehicle. It carries out path tracking control during movement. The output of the tracking controller is the local reference vector which is an input to the velocity controller. Before the movement process, $PATHCONT$ plans the path and during movement, it controls the path tracking of the vehicle. A buffer coupled to $PATHCONT$ holds the latest position and velocity estimate from the forward kinematics. This buffer communicates these estimates to $PATHCONT$.

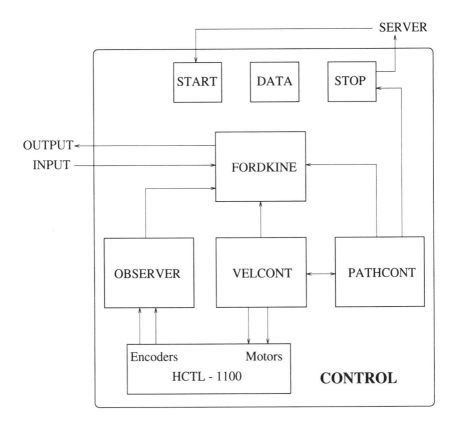

**FIGURE 6.22**
**WMR Modular Software: Nodal *CONTROL* Process**

- *VELCONT*. This process carries out the local velocity control. It contains the inverse kinematics and uses the reference body velocity vector computed by *PATHCONT* as an input. This then produces the demanded wheel parameters for the actuators at local wheel node. This is fed into the low level controller. A buffer coupled to *VELCONT* holds the latest velocity estimate from the forward kinematics. This buffer communicates this estimate to *VELCONT*.

- *OBSERVER*. This process reads the local wheel parameters from the HCTL-1100 controller card and it computes a drive velocity from the observed drive position. It outputs the observed wheel parameters to a local observer buffer process. This buffer outputs the latest wheel parameter observations to forward kinematics, *FORDKINE*.

- *FORDKINE*. This process contains the extended Information filter, EIF. It estimates the WMR motion and computes local information contributions at every time step, which are communicated to output process, *OUTPUT*.

- *STOP*. This process coordinates the termination of the local processes, when the path tracking controller has accomplished its function.

- *DATA*. The purpose of this process is to log and collect data from any active unit for use in performance analysis and validation.

## 6.6.2   Decentralized Motor Control

At the lowest level in the decentralized WMR control system is motor control. The input to this motor controller is $\mathbf{z}_{r_i}(k)$, the demanded wheel parameters from the body velocity controller, *VELCONT*. This reference signal is applied to the actuators as a pulse width modulated (PWM) voltage, feeding back observations of wheel parameters to provide control. An HCTL-1100 controller chip is used to interface the motor controller to the actuators [52]. This occurs independently and locally at each DSU module.

The Hewlett-Packard HCTL-1100 chip is a high performance, general purpose, digital motion control integrated circuit which can be used in microprocessor based, digital closed loop systems [33], [52]. It is designed to communicate with a host microprocessor through an eight bit parallel I/O port. Commands from the host processor are used by the HCTL-1100 to control an amplifier and motor combination. A motor with an incremental shaft encoder, a Hewlett-Packard HEDS-5500 encoder, is connected to the amplifier. The encoder's quadrature output is then fed back into the HCTL-1100 to close the loop [53]. Figure 6.23 shows a closed loop motion control system using the HCTL-1100 controller board.

The motor velocity control mode is used if a wheel drives under actuation. If it rotates under actuation, the steer motor position control mode is used. The sensor on the drive actuator is an encoder which makes an angle measurement. The HCTL-1100 internally computes a drive velocity, which is implemented by proportional velocity control. The steer motor has an encoder measuring wheel steer angle and is controlled using the positional control mode on the HCTL-1100. This mode implements a proportional, integral and derivative (PID) controller. In this way, each DSU decentrally achieves both drive and steer motor actuation.

## 6.6.3   WMR Trajectory Generation

The paths of the WMR are created by specifying 2D planar trajectories. Generalized 2D motion is defined by specifying three motion vectors:

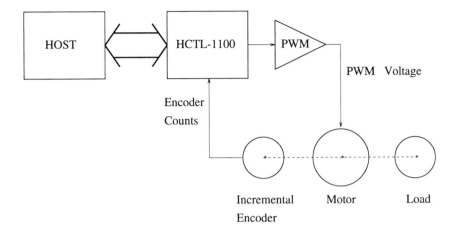

**FIGURE 6.23**
**Closed Loop Motion Control with the HCTL-1100**

position, velocity and acceleration. The three sets of motion equations, position, velocity and acceleration are sufficient to describe any planar motion of the WMR. Three main types of trajectories are generated, straight line, sinusoidal and circular (constant and increasing radius). Combinations of these are also considered, including reverse motion in any of the trajectories. Figure 6.24 shows the time profiles of position, velocity and acceleration magnitudes in straight line motion.

The motion angles are shown in Figure 6.25. From this figure, the difference between the position and velocity angles is zero. This means the difference between the vehicle orientation and heading is a constant. Figures 6.26 and 6.27 show motion magnitudes and angles for a sinusoidal trajectory. Figure 6.27 illustrates that the difference between the position and velocity angles vary with time. Circular motion is illustrated in Figures 6.28 (motion magnitudes) and 6.29 (motion angles). From Figure 6.29 the difference between the velocity and position angles is ninety degrees. Figure 6.30 shows a composite trajectory which consists of initial motion in a straight line at 45 degrees in the global frame, then motion in a circle of increasing radius and then sinusoidal motion parallel to the global frame y-axis. Results for the WMR trajectory following are presented and analyzed in Chapter 7.

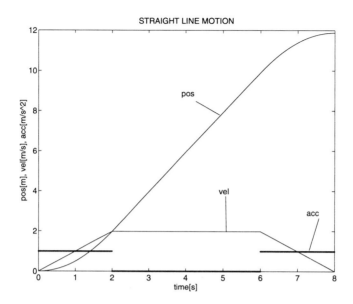

**FIGURE 6.24**
**Linear Trajectory: Position, Velocity and Acceleration Magnitudes**

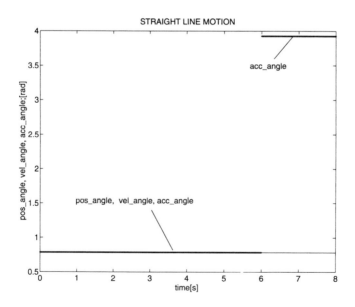

**FIGURE 6.25**
**Linear Trajectory: Position, Velocity and Acceleration Angles**

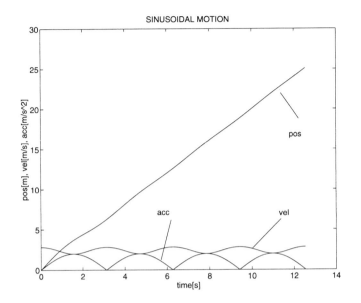

**FIGURE 6.26**
**Motion Magnitudes in a Sinusoidal Trajectory**

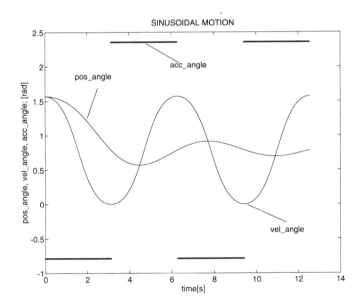

**FIGURE 6.27**
**Motion Angles in a Sinusoidal Trajectory**

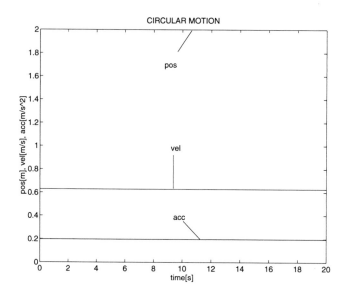

**FIGURE 6.28**
**Circular Position, Velocity and Acceleration Magnitudes**

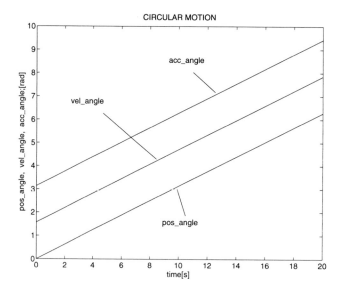

**FIGURE 6.29**
**Circular Position, Velocity and Acceleration Angles**

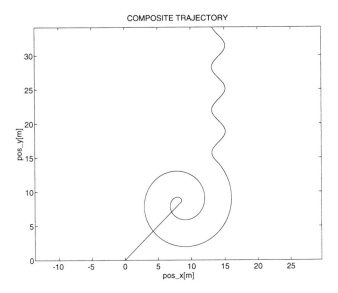

**FIGURE 6.30**
**A Composite Trajectory**

## 6.7  Summary

In this chapter the hardware and software implementation of the theory developed has been described. A general decentralized and modular kinematic model was established for a WMR with simple wheels. Specific WMR system models used were presented. The decentralized closed loop WMR control was discussed for both fully connected and non-fully connected topologies of three DSUs. The individual modules of the WMR were described and different configurations of these units built into WMR vehicles were presented. The software developed was then outlined and illustrated using pseudocode. Modular software development and functioning were explained in terms of a configuration program, specific to a particular WMR configuration, and a generic nodal program which is the same for all modules. A sample of planar motion trajectories generated was outlined.

# Chapter 7

## Results and Performance Analysis

### 7.1 Introduction

In this chapter experimental results are presented and analyzed. The main objective is to show that the proposed decentralized algorithms perform as well as equivalent centralized algorithms, thus confirming the theoretical (algebraic) equivalence between the two types of algorithms explicit in the derivation of the decentralized algorithms from centralized ones. The second objective is to demonstrate the benefits of scalable decentralized estimation and control. These two goals are achieved by presenting and analyzing results from both simulations and the WMR application. To start with, criteria for performance evaluation and validation are introduced. Results are then presented starting with those from simulations. Innovations, information (and state) estimates, tracking properties and control vectors are discussed. Real data from the WMR implementation is then presented. Here, the performance of the WMR in executing different test paths is evaluated. Wheel velocities, steer angles, vehicle orientation, wheel positions and WMR planar positions, all under decentralized kinematic control, are analyzed. The main aim in this chapter is to show that given a good centralized estimation or control algorithm an equally good decentralized equivalent can be provided.

### 7.2 System Performance Criteria

The criteria for evaluating performance are the same for both decentralized and centralized systems. System validation is carried out by comparing results from the decentralized algorithms to those obtained from conventional centralized systems.

## 7.2.1  Estimation Criteria

The criteria used to evaluate estimation performance are discussed in Chapter 2: unbiasedness, efficiency and consistency. The problem with these criteria is that they are theoretical concepts which depend on knowledge of the *true* state. In practical systems the true state is never known; that is the reason why estimation is applied in the first place. The most practical estimation performance criteria is investigating the properties of the *innovations*, from which properties of the state estimates can then be inferred. The innovation is the difference between the actual observation and predicted observations. The innovation vector is given by

$$\nu(k) = \mathbf{z}(k) - \mathbf{H}(k)\hat{\mathbf{x}}(k \mid k - 1)$$
$$\nu(k) = \mathbf{z}(k) - \mathbf{h}(\hat{\mathbf{x}}(k \mid k - 1)),$$

for linear and nonlinear systems, respectively. It is the most important and practical measure of estimator performance because it gives the deviation between filter estimated observations and true observation sequences. The innovation covariance $\mathbf{S}(k)$ is a measure of the estimated innovation deviations.

In this book three properties of the innovation are used to judge estimation performance.

- **Magnitude of the innovations**

  95% of the innovations must lie within the $2\sigma$-gate. The $2\sigma$-gate is the 95% confidence bound for a Gaussian random variable. This is the simplest test for filter correctness and is carried out by looking directly at the innovation sequence. If too many innovations fall outside this gate, this indicates that the noise levels set in the filter are generally lower than the true noise levels. For example, a filter whose innovation sequence has 40% of its innovations *outside* the region between $-2\sigma$ and $2\sigma$ is said to be *mismatched*. Mismatching is a manifestation of estimation inconsistency.

- **$\chi^2$ test of the normalized innovation squared**

  The innovation should be Gaussian with zero mean (unbiased) with computed covariance $\mathbf{S}(k)$. This is tested by performing a $\chi^2$ test on the normalized innovations squared. If the overall noise level is not suspect, a simple computation of the sequence mean is sufficient to show that the innovations are unbiased.

- **The autocorrelation of the innovation sequence**

  The innovations should be white, that is, uncorrelated in time. This is tested by computing the autocorrelation function of the innovation sequence.

Usually the three innovation tests are required to provide rigorous analysis of filter performance. For the work presented in this book, however, it is considered sufficient to use only the magnitude test to evaluate the performance and properties of the estimators for both simulations and real data. This is done by plotting the innovation sequences against time and checking whether the innovations lie within the 95% confidence bound. For simulations, state and information estimates are also plotted and analyzed.

## 7.2.2 Control Criteria

In this book the control performance is judged by comparing the estimated states to the reference states for the implemented WMR decentralized closed loop velocity control. This is done for the steer angle and velocity of each DSU wheel in three different trajectories. Although the DSUs do not control the WMR body state vector, they compute an estimate of the body state vector. The errors between WMR body velocity and steer angle estimates and the reference steer angle and velocity are similarly analyzed for each DSU. Errors in estimated vehicle orientation and rate of change of orientation from the referenced ones are considered.

In this analysis the difference between the state estimate and the reference state at discrete time step, $k$, is defined as the estimated control error, $e(k)$. This error is plotted against time for the different trajectories. The data are quantitatively analyzed by computing and tabulating three analytical measures defined below.

- **Mean Error**
  The mean error $\bar{e}$ is defined as the average error from the demanded state,

$$\bar{e} = \frac{1}{N} \sum_{j=1}^{N} e(k) \qquad (7.1)$$

  where $e(k)$ is the estimated control error and $N$ is the number of time steps. This quantifies any systematic offset from the reference state.

- **Estimated Standard Deviation**
  The estimated standard deviation of the errors from the mean error $\hat{\sigma}$ is given by

$$\hat{\sigma} = \sqrt{\frac{1}{N} \sum_{j=1}^{N} (e(k) - \bar{e})^2}. \qquad (7.2)$$

  This is used to quantify the spread of the error, ignoring the effects of any systematic offset from the demanded state.

- **Root Mean Square Error**
  The root mean squared error $e_{rms}$ is defined by

$$e_{rms} = \sqrt{\frac{1}{N} \sum_{j=1}^{N} e^2(k)}. \qquad (7.3)$$

  It is a measure of the absolute size of the error. If the systematic error is zero, then this is equivalent to $\hat{\sigma}$.

## 7.3  Simulation Results

Validation is carried out by comparing estimation and control results from three types of systems: conventional centralized (C), fully connected decentralized (D) and non-fully connected decentralized (DD). This is done by plotting and analyzing results for the same system state $x_1(k)$ (the position of mass 1 in Figure 4.1), using the three different approaches: C, D, DD. The system models are given in Section 5.4. Within the non-fully connected network, results from different nodes about the same system state are compared to establish whether the network produces correct and consistent results. This is done for four arbitrary nodes in a network of four non-fully connected nodes, described in Section 5.4. For the fully connected case, the estimates and control vectors at each node are compared.

### 7.3.1  Innovations

Figure 7.1 shows the innovations for the single state $x_1(k)$ for the three systems: C, D and DD. In terms of validation the results are identical and hence indistinguishable (just one curve). As for filter performance, in all three cases only 3 out of 100 points (3%) are outside the $2\sigma$ confidence region, thus satisfying the 95% confidence rule. This implies that the filter is consistent and well-matched. By inspection and computing the sequence mean, the innovations are shown to be zero mean with variance $\mathbf{S}(k)$. Practically it means the noise level in the filter is of the same order as the true system noise. There is no visible correlation of the innovations sequences. This implies that there are no significant higher-order unmodeled dynamics nor excessive observation noise to process noise ratio.

Figure 7.2 shows innovations for state $x_1(k)$ obtained from four nodes in a non-fully connected network of control nodes. The innovations at nodes 0, 1, and 2 are identical and satisfy the 95% confidence rule. This shows that filter is consistent and well-matched. The sequences are unbiased and uncorrelated. At node 3 the innovation is zero indicating that state $x_1(k)$ is irrelevant to node 3 and hence it is neither estimated nor controlled. This follows from the model distribution models in Section 5.4. If the network is fully connected, the innovation vector obtained at each node is the same for all nodes and is identical to the one obtained in an equivalent centralized system.

### 7.3.2   State Estimates

Figure 7.3 compares state estimates from the three methods: C, D and DD, for the particular state $x_1(k)$ with an applied reference signal $x_{r1}(k)$. Again the estimates curves are identical. For all three methods the reference, a straight line curve, lies in the middle and the estimate is always well placed between the observation and state prediction. This means there is balanced confidence in observations and predictions. Since, as $k \to \infty$, the process noise variance $\mathbf{Q}(k)$ governs the confidence in predictions and the observation noise variance $\mathbf{R}(k)$ governs the confidence in observations, the results are an indication that the noise variances are well chosen. As a result, no more confidence than is justified is granted to estimates.

In Figure 7.4 state estimates in a scalable decentralized control network are compared, by considering estimation of the single state $x_1(k)$. The estimates at nodes 0, 1 and 2 are identical and their characteristics are as discussed above. The estimate at node 3 is zero showing that the state $x_1(k)$ is irrelevant to this node.

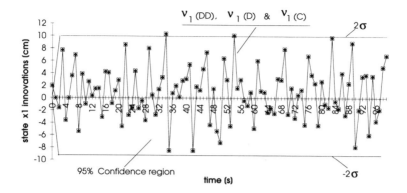

**FIGURE 7.1**
**Innovations: Distributed Decentralized, Decentralized and Centralized**

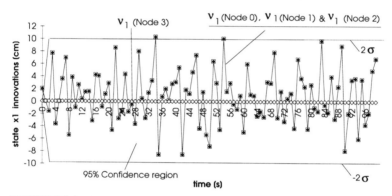

**FIGURE 7.2**
**Innovations: Scalable Decentralized Control Network**

### 7.3.3    Information Estimates and Control

Figure 7.5 compares information estimates for the three cases, C, D and DD. As with innovations and state estimates, the information estimates are identical. The estimates closely track the reference information signal without significant lead or lag. Comparison of information estimation within a scalable control network is as for state estimates, with no information being estimated by node 3. This is demonstrated in Figure 7.6. By comparing the curves of the information estimates and those of the state estimates, it is shown that the information estimates are scaled state estimates. This is because the elements of the information matrix $\mathbf{Y}(k \mid k)$ settle at constant values. As discussed in Chapter 5, the control vector $\mathbf{u}(k)$ depends on the estimated control error $\mathbf{e}(k)$ (the difference between the state estimate

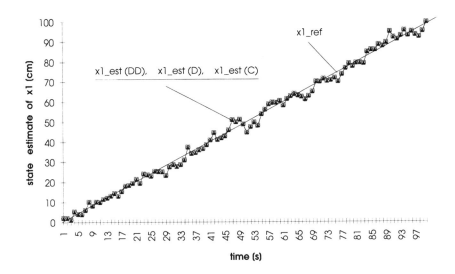

**FIGURE 7.3**
**State Estimate: Distributed Decentralized, Decentralized and Centralized**

and reference vectors) and the Backward Riccati control (law) gain $\mathbf{G}(k)$. Considering an individual state $x_1(k)$, its control signal $u_1(k)$ is illustrated for the three control systems in Figure 7.7. The corresponding control law (gain) is exactly the same for all three systems. This control law guarantees optimal reference state tracking and efficient filtering.

Figure 7.8 compares control signals of the same state $x_1(k)$ in a network of non-fully connected control nodes. The properties of these signals and their corresponding control gains are described above. Where the state $x_1(k)$ is relevant (nodes 0, 1 and 2), its control is the same, and where it is irrelevant (node 3), the state is neither estimated nor controlled. When the topology is fully connected, the control vector is the same at every node and identical to the one obtained in a centralized system.

## 7.4  WMR Experimental Results

The application of decentralized estimation and control to the WMR is tested by using trajectory path following. The tracking is done under closed loop decentralized control. The trajectories are generated as discussed in

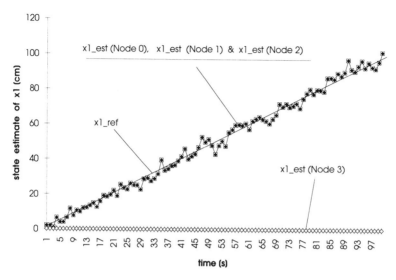

**FIGURE 7.4**
**State Estimate: Scalable Decentralized Control Network**

Chapter 6. Three general types of 2D planar paths are experimented with: linear, circular and sinusoidal. The WMR has three DSUs and 3DOFs. The implementation is discussed by first presenting the motion of the three driven and steered wheels and the WMR body motion as it follows different specified paths. The performance is then analyzed by presenting and discussing the local WMR innovations and estimated control errors.

## 7.4.1    Trajectory Tracking

### Straight Line Motion

Several test runs are performed with the WMR moving in a straight line. Figure 7.14 shows the velocity profile of the individual WMR wheels in forward and reverse straight line motion. The vehicle accelerates constantly from rest to a maximum velocity, moves at this constant velocity and then decelerates constantly via rest, into reverse motion. It then reaches a maximum reverse velocity, moves at this velocity and then decelerates to rest. This is achieved by locally actuating the DSU wheel using the driving motor. The steer angle is zero. The orientation and heading of the virtual wheel $b$ are the same. Figure 7.15 shows the variation of the DSU wheel position from the origin as a function of time. The WMR body moving in a straight line is shown in Figure 7.9. The actual dimensions of the vehicle are used in this representation.

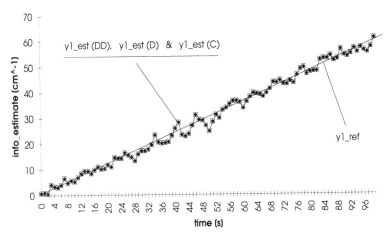

**FIGURE 7.5**
**Information Estimate: Distributed Decentralized, Decentralized and Centralized**

**Sinusoidal Motion**

Figure 7.16 shows wheel steer angle variation in sinusoidal motion. The difference between heading and orientation varies continuously with time. The magnitude of wheel velocity is kept constant as illustrated in Figure 7.17. The motion of the WMR body frame in sinusoidal motion is shown in Figure 7.10. The WMR performs both forward and reverse motion in this sinusoidal pattern.

**Circular Motion**

The WMR is subjected to several variations of circular trajectory. Circular motion with a constant radius is achieved by driving each wheel at a constant velocity with a linearly increasing steer angle, as illustrated in Figure 7.18. Figure 7.19 shows the x and y positions of the wheels in the floor (global) frame. The difference between the heading and orientation of the virtual wheel is $\pi/2$ radians. Figure 7.11 shows the WMR body frame in circular motion. The vehicle can move forward and backwards in circular motion, in any planar direction irrespective of its posture. Circular motion which describes a "figure of eight" is also used.

Semi-circular motion is practically convenient because it allows the motor cables to be wound and unwound in motion without having to stop the vehicle. The motion is implemented by prescribing a planar semi-circular trajectory. The WMR body frame describing this semi-circular motion is shown in Figure 7.12. The vehicle can reverse back and forth in this trajectory. The reverse motion can be carried out in any direction while

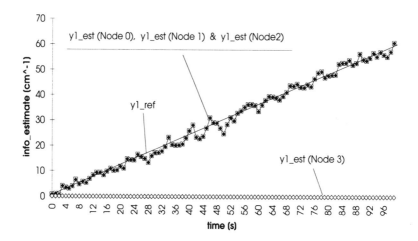

**FIGURE 7.6**
**Information Estimate: Scalable Decentralized Control Network**

describing part of a "figure of eight". Figure 7.13 shows the WMR body switching between simple circular motion and reversal motion in a "figure of eight". The wheel velocities, steer angles are independently controlled at each DSU.

## 7.4.2   Innovations and Estimated Control Errors

Estimation performance of the WMR is evaluated by presenting and analyzing a sample of the innovations of DSU wheel parameters. Figure 7.21 shows DSU 1 wheel velocity innovations for the vehicle moving in a "figure of eight". The corresponding vehicle trajectory is shown in Figure 7.12. Figure 7.20 shows the innovations of the wheel steer angle of DSU 1 where the vehicle is moving in circular motion. The effectiveness of control is analyzed by presenting the estimated control errors of the DSU wheels and that of the WMR body. Table 7.1 shows the analysis of these errors for the three DSUs and the virtual wheel when the vehicle is moving in a straight line. Similarly, Tables 7.2 and 7.3 show the analysis for circular and sinusoidal motion. A sample of the estimated control error plots is presented. Figure 7.22 shows the velocity estimated control error for DSU 3 when the WMR is in linear motion. Figure 7.23 shows the estimated control error of the DSU 2 steer angle when the vehicle is moving in sinusoidal motion.

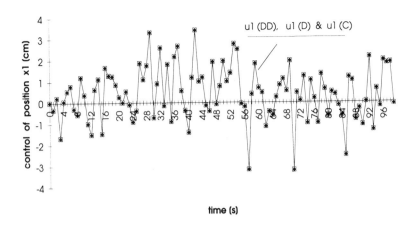

**FIGURE 7.7**
**Control: Distributed Decentralized, Decentralized and Central-ized**

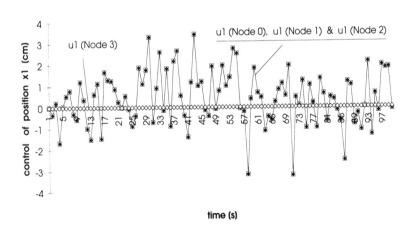

**FIGURE 7.8**
**Control: Scalable Decentralized Control Network**

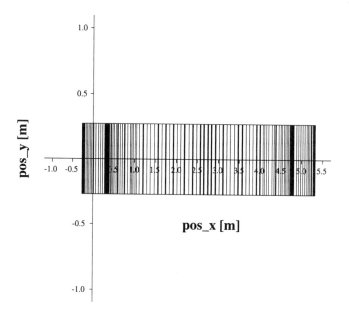

**FIGURE 7.9**
**WMR in Straight Line Motion**

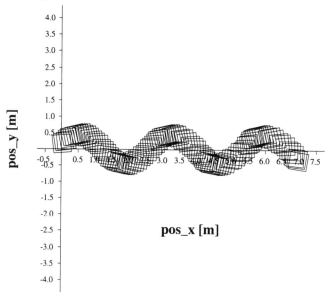

**FIGURE 7.10**
**WMR in Sinusoidal Motion**

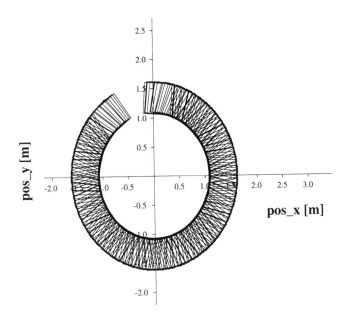

**FIGURE 7.11**
**WMR in Circular Motion**

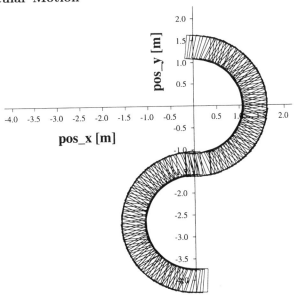

**FIGURE 7.12**
**WMR in Semi-Circular Motion**

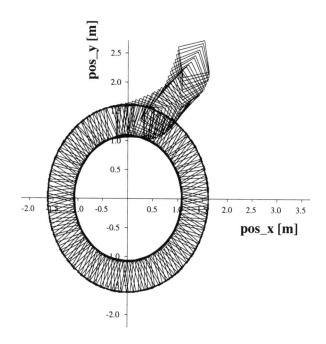

**FIGURE 7.13**
**WMR in Circular Motion with Reverse**

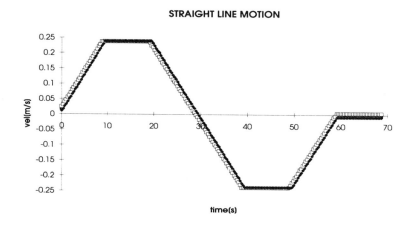

**FIGURE 7.14**
**Forward and Reverse Linear Motion: Wheel Velocity Profile**

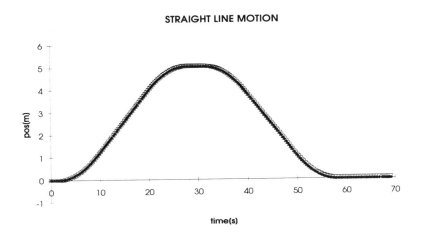

**FIGURE 7.15**
**Wheel Position in Forward and Reverse Linear Motion**

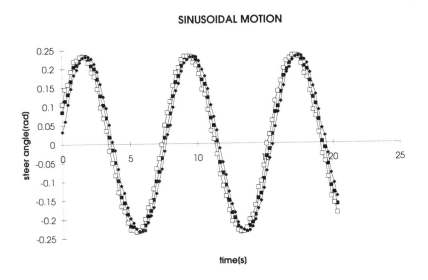

**FIGURE 7.16**
**Wheel Steer Angle in Sinusoidal Motion**

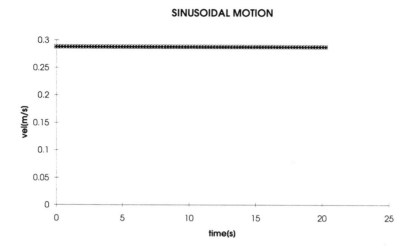

**FIGURE 7.17**
**Wheel Velocity Magnitude in Sinusoidal Motion**

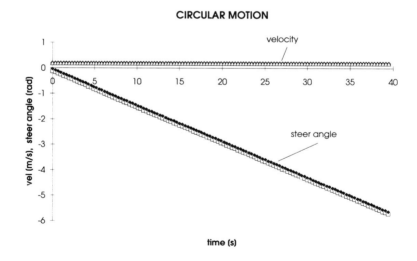

**FIGURE 7.18**
**Circular Motion: Wheel Steer Angle and Velocity Magnitude**

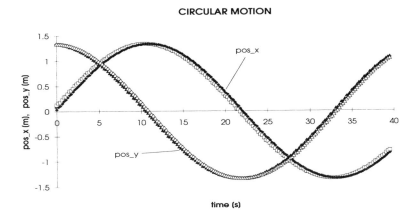

**FIGURE 7.19**
**Circular Motion: Wheel x-y Positions**

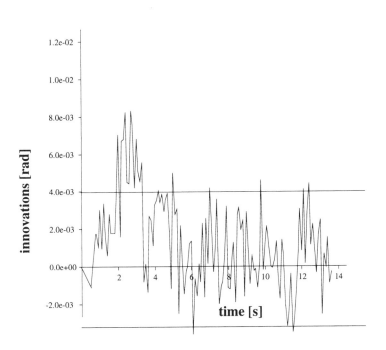

**FIGURE 7.20**
**DSU 1 Wheel Steer Angle Innovations**

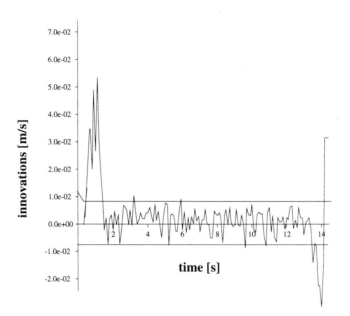

**FIGURE 7.21**
**DSU 1 Wheel Velocity Innovations**

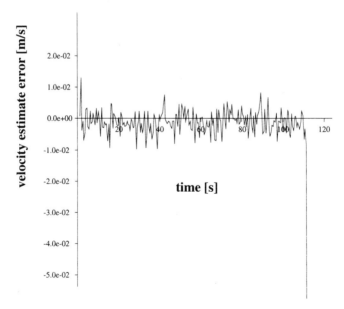

**FIGURE 7.22**
**DSU 3 Wheel Velocity Estimated Control Error**

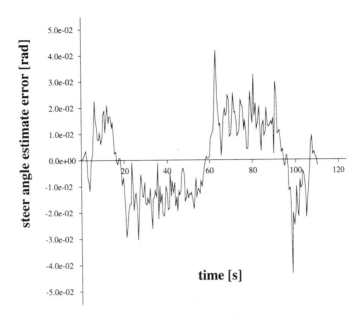

**FIGURE 7.23**
**DSU 2 Wheel Steer Angle Estimated Control Error**

Each DSU has the capacity to estimate the WMR body parameters. Errors between these estimates and demanded WMR body states are included in Tables 7.1, 7.2 and 7.3. Figure 7.24 shows the virtual wheel steer angle error obtained at DSU 1. In Figure 7.25 this estimate is compared with the one from DSU 2. Figure 7.26 shows the estimated control error of the rate of change of body orientation computed when the WMR is pursuing a sinusoidal trajectory. Figure 7.27 shows the orientation error obtained by integration when the vehicle is moving with constant orientation in a straight line. The control gains cost functions are shown in Tables 7.4, 7.5 and 7.6.

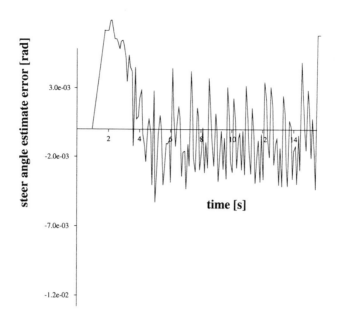

**FIGURE 7.24**
**WMR Body Steer Angle Estimated Control Error: DSU 1**

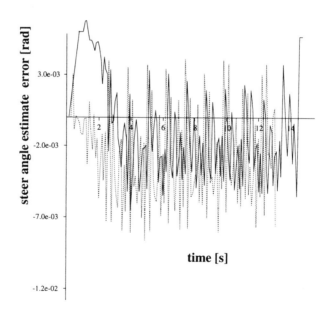

**FIGURE 7.25**
**WMR Body Steer Angle Estimated Control Error: DSU 1 and 2**

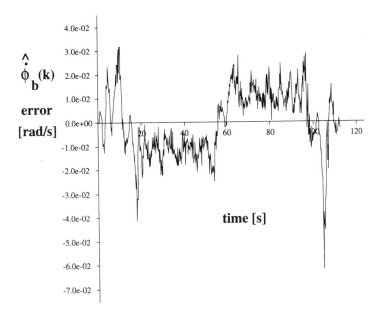

**FIGURE 7.26**

**WMR Body $\dot{\phi}_b(k)$ Estimated Control Error: DSU 1**

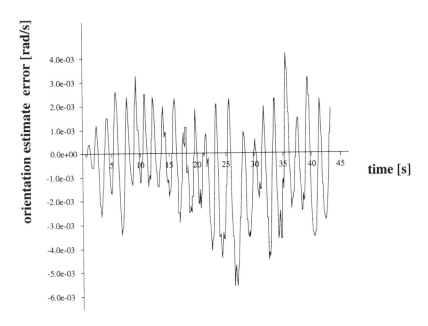

**FIGURE 7.27**
**WMR Body Orientation Estimated Control Error: DSU 3**

**Table 7.1**  Analysis of WMR Body Estimated Control Errors
in Linear Motion

| WMR wheel | 0 | 1 | 2 | $b$ |
|---|---|---|---|---|
| $\bar{e}_V$ | 0.0017 | 0.0015 | -0.0004 | -0.003 |
| $\sigma_V$ | 0.0008 | 0.0007 | 0.0006 | 0.0012 |
| $e_{rmsV}$ | 0.0015 | 0.0020 | 0.0018 | 0.0039 |
| $\bar{e}_\gamma$ | 0.0015 | 0.0020 | 0.0020 | -0.0027 |
| $\sigma_\gamma$ | 0.0007 | 0.0008 | 0.0010 | 0.0011 |
| $e_{rms\gamma}$ | 0.0030 | 0.0027 | 0.0034 | 0.0047 |

**Table 7.2**  Analysis of WMR Body Estimated Control Errors
in Circular Motion

| WMR wheel | 0 | 1 | 2 | $b$ |
|---|---|---|---|---|
| $\bar{e}_V$ | 0.0027 | 0.0025 | 0.0030 | 0.0060 |
| $\sigma_V$ | 0.0010 | 0.0010 | 0.0016 | 0.0012 |
| $e_{rmsV}$ | 0.0030 | 0.0028 | 0.0029 | 0.0039 |
| $\bar{e}_\gamma$ | 0.0015 | 0.0020 | 0.0020 | 0.0027 |
| $\sigma_\gamma$ | 0.0009 | 0.0018 | 0.0018 | 0.0020 |
| $e_{rms\gamma}$ | 0.0030 | 0.0027 | 0.0034 | 0.0047 |

## 7.5  Discussion of Results

### 7.5.1  Local DSU Innovations

A sample of the innovation results in Figures 7.21 and 7.20 show good
estimation performance. In both cases the magnitudes of the innovations
fall within the the $2\sigma$ gates except at the beginning. This means the filter
noise levels were set close to the true wheel noise levels. In both cases
there is an initial spike of innovations which lie outside the bound, but
the sequences quickly settle. The initial spike is due to unmodeled wheel
inertias. There is no visible bias nor correlation of the innovation sequences.
It can be deduced that the system does not have any significant high order
unmodeled dynamics.

**Table 7.3** Analysis of WMR Body Estimated Control Errors in Sinusoidal Motion

| WMR wheel | 0 | 1 | 2 | $b$ |
|---|---|---|---|---|
| $\bar{e}_V$ | 0.0024 | 0.0025 | 0.0024 | 0.0043 |
| $\sigma_V$ | 0.0018 | 0.0015 | 0.0015 | 0.0020 |
| $e_{rmsV}$ | 0.0065 | 0.0070 | 0.0068 | 0.0089 |
| $\bar{e}_\gamma$ | 0.0018 | 0.0022 | 0.0020 | 0.0020 |
| $\sigma_\gamma$ | 0.0010 | 0.0005 | 0.0009 | 0.008 |
| $e_{rms\gamma}$ | 0.0030 | 0.0060 | 0.0054 | 0.0090 |

**Table 7.4** Control Gains and Constants for Linear Motion

| WMR DSU | 0 | 1 | 2 |
|---|---|---|---|
| $\mathbf{G}_V(k)$ | 0.55 | 0.70 | 0.65 |
| $\mathbf{G}_\gamma(k)$ | 0.81 | 0.80 | 0.75 |
| $\mathbf{X}_V(k)$ | 2.0 | 2.0 | 2.0 |
| $\mathbf{X}_\gamma(k)$ | 80.0 | 80.0 | 80.0 |
| $\mathbf{U}_V(k)$ | 40.0 | 40.0 | 40.0 |
| $\mathbf{U}_\gamma(k)$ | 2.0 | 2.0 | 2.0 |

**Table 7.5** Control Gains and Constants for Circular Motion

| WMR DSU | 0 | 1 | 2 |
|---|---|---|---|
| $\mathbf{G}_V(k)$ | 0.76 | 0.70 | 0.85 |
| $\mathbf{G}_\gamma(k)$ | 0.53 | 0.35 | 0.40 |
| $\mathbf{X}_V(k)$ | 100.0 | 100.0 | 100.0 |
| $\mathbf{X}_\gamma(k)$ | 100.0 | 100.0 | 100.0 |
| $\mathbf{U}_V(k)$ | 3.0 | 3.0 | 3.0 |
| $\mathbf{U}_\gamma(k)$ | 3.0 | 3.0 | 3.0 |

**Table 7.6**  Control Gains and Constants for Sinusoidal Motion

| WMR DSU | 0 | 1 | 2 |
|---|---|---|---|
| $\mathbf{G}_V(k)$ | 0.85 | 0.9 | 0.95 |
| $\mathbf{G}_\gamma(k)$ | 0.5 | 0.35 | 0.60 |
| $\mathbf{X}_V(k)$ | 5.5 | 5.5 | 5.5 |
| $\mathbf{X}_\gamma(k)$ | 150.0 | 150.0 | 150.0 |
| $\mathbf{U}_V(k)$ | 50.0 | 50.0 | 50.0 |
| $\mathbf{U}_\gamma(k)$ | 3.0 | 3.0 | 3.0 |

## 7.5.2   Wheel Estimated Control Errors

The complete analysis of the DSU wheel estimated control errors is shown in Tables 7.1, 7.2 and 7.3. In most of the trajectories the DSU proved capable of effective tracking of reference velocity and steer profiles. Figure 7.22 shows a typical estimated control error in wheel velocity. The curve has a slight bias ($\bar{e} = -0.0005$). This illustrates effective control of the difference between the steer angle estimate and its reference. The complete analysis of the DSU wheel estimated control errors is shown in Tables 7.1, 7.2 and 7.3. There are large root mean square errors, $e_{rms}$ for sinusoidal motion and circular motion. This is because when the WMR tries to achieve large changes in path curvature, large errors occur. Figure 7.23 illustrates this trend for a steer angle estimated control error, as the WMR executes sinusoidal motion. The error mean is a sinusoid. This is due to motion slip in wheels as the vehicle turns around corners.

Figure 7.26 shows the estimated error in the rate of change of vehicle orientation as the vehicle moves in a "figure of eight". As the vehicle turns, the mean of the error shifts up and down the time axis. The local reference velocity and steer angle profiles are correlated to the curvature of the reference path. This is illustrated by comparing the virtual steer angle and velocity profiles to the referenced trajectories in Chapter 6.

## 7.5.3   WMR Body Estimates

From the error analysis in Tables 7.1, 7.2 and 7.3 the errors are larger in the virtual wheel than in the physical wheels. This is because the body estimates are not directly controlled; they are only computed from DSU estimates through the forward kinematics. The body estimated control errors show correlation with the change in heading. The body angular velocity is calculated from the virtual wheel steer angle, hence these two are highly correlated. Each DSU is capable of producing estimates of the body motion.

Figure 7.25 compares the WMR body steer estimated control error as computed by DSU 1 and 2. There is initially a large difference between estimates followed by steady delay. This is due to time delays in computation and communication. In circular motion shown in Figure 7.11, when the final position of the WMR was physically measured, it was found that the offset from the goal location was much larger than that estimated. This is because the position estimate is computed by integrating the body velocity obtained from the local DSU estimates. As a result the uncertainty associated with the position increases as the WMR moves (as in a random walk). The location of the vehicle may not be the goal location, due to unmodeled process noise (slip). In the case of circular motion, the vehicle tends to describe circles of larger radii than that prescribed.

## 7.6 Summary

The results obtained from implementing the theory developed in this book have been presented and discussed. Decentralized algorithms were shown to produce the same results as centralized systems, which confirms the theoretical equivalence between the two types of algorithms. By employing estimation and control performance criteria the algorithms are shown to exhibit good filtering and control characteristics. Specifically decentralized estimation, both fully connected and non-fully connected, are shown to be efficient and well matched. The limitations of full connectedness and their resolution by use of model distribution and local internodal communication are shown by comparing results from two networks, one fully connected and the other non-fully connected (model defined topology). In this way the benefits of model distribution in decentralized systems are demonstrated.

The WMR experimental results show that decentralized estimation and control can be implemented for a mobile vehicle using modular software and distributed hardware. The three driven and steered units communicate to produce smooth trajectories. The way the vehicle tracks different trajectories, while each DSU wheel was locally driven and steered, shows effective modular motor actuation, given local encoder odometry information and inter-module communication. The results demonstrate that the WMR can execute omni-directional trajectories under local closed loop control. Local control gains generated by the Backward Riccati equations showed effective body velocity control through local wheel control. However, integrating information from local velocity control results in unmodeled errors. For accurate path tracking the vehicle requires a sensor accurately measuring vehicle position. The DSU integrated velocity data is useful, if it is complimented by a supply of positional updates.

# Chapter 8

## Conclusions and Future Research

### 8.1 Introduction

This chapter draws conclusions from the material presented in the book and discusses possible directions for future work. It summarizes the developed theory and the experimental results, as well as evaluate the contributions made. Specific focus is on the benefits of using decentralized estimation and control, model defined topologies and local internodal communication. The chapter also discusses the application of this theory to modular vehicle robotics. Further, it explores unresolved theoretical, design and application issues. It is within this context that suggestions for further research are made. At the theoretical level, recommendations are made to expand and further generalize the proposed estimation, communication and control ideas. Proposals are also made for the extension and application of the concepts developed in this book to other fields of engineering and scientific research.

### 8.2 Summary of Contributions

This book has made significant contributions to the development of decentralized estimation and control theory. In particular, an information space approach has been employed to produce robust, flexible and scalable decentralized data fusion and control algorithms for multisensor and multiactuator systems. This work has been used as a basis for the design, construction and control of a modular wheeled mobile robot.

### 8.2.1  Decentralized Estimation

The linear Information filter was generalized to deal with the problem of estimation of nonlinear systems. This was done by deriving the extended Information filter (**EIF**). For multisensor and multiactuator systems, the filter was decentralized to give the decentralized extended Information filter (**DEIF**). This is a powerful estimation technique which, soon after its publication, found applications in independent work by other researchers [33], [70], [104], [118]. *Generalized* internodal information transformation theory was proposed as a solution to the drawbacks of fully connected decentralization. As a result, non-fully connected topologies which are model defined were developed for both estimation and control problems. Special cases of the transformation matrices were derived and both their practical implications and applications were identified.

For linear systems a distributed and decentralized Information filter (**DD-IF**), which is more general than the one in [20], was derived. For nonlinear systems the distributed and decentralized extended Information filter (**DDEIF**) was developed. This is a non-fully connected decentralized data fusion algorithm with local reduced order models and minimized communication. The novelty of this algorithm is that it can be applied to nonlinear systems and that it does not require full connection of nodes or propagation of information between unconnected nodes. This distinguishes it from the fusion topologies in the literature [46], [54], [115].

### 8.2.2  Decentralized Control

In extending these decentralized data fusion systems to stochastic control problems, decentralized sensor based control strategies were proposed for both linear and nonlinear systems. The decentralized Information filter was extended to a decentralized linear control system and the decentralized extended Information filter to a decentralized control algorithm for nonlinear systems. To reduce the problems associated with fully connected systems, the distributed and decentralized Information filter was extended to distributed and decentralized linear control. Similarly, the distributed and decentralized extended information filter was extended to distributed and decentralized control algorithms that handle nonlinearities. This is the major *composite* theoretical result from this book. It provides a scalable, flexible and robust architecture for nonlinear related estimation and control problems.

### 8.2.3  Applications

The theoretical results were demonstrated on Transputer based hardware in the form of a modular navigating wheeled mobile robot (WMR). In the process a decentralized kinematic model and modular software for a general WMR with simple wheels were produced. The result showed that the design, construction and control of a completely modular WMR is feasible. Several configurations of the vehicle were built as part of Oxford Navigator project (OxNav). The work presented in this book formalizes the principles of its design, construction and operation. The theory allows the nonlinear WMR data fusion and control functions to be fully decentralized and distributed. The advantages of the modular vehicle include scalability, application flexibility, low prototyping costs and high reliability. From the experimental results, the algorithms tested on the WMR satisfied standard estimation and control performance criteria. This showed that given a good centralized estimator or controller, an equivalent decentralized system of equal performance can be obtained. In addition, decentralization provides functional advantages for the system.

The scope of application of the theoretical contributions made goes beyond vehicle robotics. The theory described finds application in many large multisensor and multiactuator systems such as smart structures, process plant control, space structures, surveillance and econometrics.

## 8.3   Research Appraisal

The relevance and limitations of the work described in this book are best appreciated in the context of existing decentralized systems research. The work fills the gaps in Large Scale Systems (LSS) theory, where decentralization has been defined only in terms of an interacting hierarchy of two or more subsystems. As an alternative paradigm, this research has proposed a fully decentralized architecture for both estimation and control systems. Unlike existing LSS algorithms, the performance criteria and results are the same for both centralized and decentralized systems. This book generalizes those systems in the literature which are already fully decentralized to provide model defined, non-fully connected topologies. The resulting scalable algorithms are applicable to nonlinear systems. Hence, the developed methods are practically useful. The problem of expensive propagation of information is resolved by model defined internodal communication.

### 8.3.1   Decentralized Estimation

While the use of information space has several advantages, which include easy filter initialization and decentralized fusion, it has some associated problems. The value of the information matrix (the inverse covariance matrix) theoretically tends to infinity for steady state conditions, as the covariance matrix goes to zero. A systematic way to deal with such singularities would be to use the Joseph form of the covariance update equation [13]. Another issue is the excessive inversion of matrices required in information based estimation. Tracking the information matrix is only computationally less demanding for systems where the dimension of the measurement vector is larger than that of the state vector.

The value of the EIF is further enhanced by its flexibility to work with recently developed techniques for improving the accuracy and generality of Kalman and extended Kalman filters. Specifically, the Unscented Transform provides a mechanism for applying nonlinear transformations to the mean and covariance estimates that is provably more accurate than standard linearization [62], [60], [105]. A matrix analogous to the Jacobian can be generated from the Unscented Transform and used instead of the Jacobians in the EKF and EIF equations, thus providing improved accuracy.

The EIF can also be extended to exploit the generality of the Covariance Intersection (CI) in order to remove the independence assumptions required by all Kalman-type update equations [61], [123], [124]. The highly restrictive requirement that process and observation noise sequences must be uncorrelated is impossible to satisfy in any nonlinear filter, which means that additional stabilizing noise must be injected to mitigate the errors resulting from assumed independence. In many cases it has been shown that the use of the CI update equations can lead to improved accuracy by eliminating the need for this stabilizing noise. The incorporation of CI into the EIF simply involves the replacement of inverse covariance matrix $\mathbf{P}^{-1}(k \mid k)$ and observation noise covariance matrix $\mathbf{R}^{-1}(k)$ with $\omega \mathbf{P}^{-1}(k \mid k)$ and $(1-\omega)\mathbf{R}^{-1}(k)$, respectively, where the value of $\omega$ is a scalar parameter defined by the CI filter to minimize a particular measure of covariance size.

For the decentralized estimation algorithms proposed in this book to give the same results as corresponding centralized algorithms they must be run at *"full rate"*, which means that communication of information has to be carried out after every measurement [35], [36]. If the frequency of communication is less than the frequency of measurement, the decentralized fusion algorithms become suboptimal with respect to equivalent centralized algorithms. With full communication rate, the decentralized estimation algorithms perform exactly the same as their centralized counterparts, which confirms the theoretical equivalence (algebraic equivalence), but have the advantages of decentralization.

Some of the applications in the book belong to a specific class of decentralized architectures in which relevant measurements are simultaneously accessible to processors while assuming synchronized sensors and perfect communication reliability. In the case of asynchronous sensors, the strategy that each node broadcast its latest estimate and each node replaces its estimate by the received estimate can be used, assuming that the estimate update, communication and reception are instantaneous, and that no two updates occur simultaneously [14]. The updates are done with time-varying sampling periods and the discrete time state equations will have to take this into account accordingly. In practice the above assumptions are not perfectly satisfied, hence the strategy yields suboptimal global estimates.

### 8.3.2 Decentralized Control

The separation principle does not strictly hold for nonlinear stochastic control problems and the use of assumed certainty equivalence might not be sufficient for certain highly nonlinear systems. The kinematic model used for the WMR is sufficient for low speed navigation whereas for higher speeds a dynamic model is required. The vehicle positional estimate is obtained by integration. Errors arising from the integration can be minimized by using information from sonar sensors to supplement this estimate.

The theoretical framework for studying general decentralized control is difficult to establish. This is because optimal decentralized control for heavily coupled large scale systems is complicated, nonlinear and involves the several controllers 'signaling' each other. Consequently, generalized decentralization tends to be impractical while from a theoretical viewpoint it is sometimes ad hoc (and hence suboptimal), whereas better control (often optimal) can always be achieved from a central controller. The decentralized control algorithms developed in this book provide optimal control if the system is either fully connected, easily decoupled (systematic model distribution) or weakly coupled (forced model distribution). In the case of weakly coupled systems the control achieved is strictly suboptimal, that is, there is a slight degradation in performance due to decentralization. In any decentralized architecture (estimation or control) there are costs and penalties associated with extensive communication, more processors, redundant computation and time delays.

## 8.4 Future Research Directions

The work described in this book can be extended by taking cognizance of the limitations of the proposed algorithms and by considering unresolved but related estimation and control issues. New research directions are postulated for both theoretical work and applications.

### 8.4.1   Theory

There are quite a number of ways of extending the theory developed in this book. First, by providing a more general definition of information in terms of entropy, the algorithms can be extended to a wider range of data fusion issues. This could be done by expanding the relationship between entropy and information, and then deriving the algorithms presented here in that framework. Such work could find applications in discrete problems. A second research direction is to extend the theory developed to lumped parameter systems theory. The results could then be applied to flexible structures, in particular, integrated actuator theory and aircraft systems.

A third direction would be to extend the work to system wide optimality problems. This is an area where the objective is to decide which sensor or actuator is most suitable for a particular function in a multi-functional system. It will also be of interest to extend the estimation and control work to robust estimation and $H^\infty$ control systems. To deal with the problem of the invalidity (in general) of the separation principle for nonlinear stochastic problems, methods of linear perturbation control (LQG direct synthesis), closed-loop controller ("dual control" approximation) and stochastic adaptive control should be considered in information space.

The notion and characteristics of nonlinear nodal transformation should be investigated further. The internodal transformation techniques could be extended to deal with transformation nonlinearities and uncertainties. In terms of system performance, rigorous stability, robustness and reliability analysis methods should be employed to evaluate the algorithms.

The Unscented Transform and Covariance Intersection methods provide a variety of performance advantages over the methods traditionally used with Kalman and extended Kalman filters [62], [61]. It will be interesting to explore their use with the EIF. All results relating to the EIF can be easily extended to exploit the benefits of the Unscented Transform and CI. Such extensions will almost surely be desirable in real-world applications of the techniques described in this book. Consequently, this is one potentially profitable research direction.

Work to establish a theoretical framework for studying general decentralized control should continue. The objective and challenge should be to obtain optimal decentralized control for heavily coupled large scale systems and complex systems. The decentralized control problem should be formulated in information space and approached directly rather than by extension from the decentralized estimation problem. The issue of robustness should also be addressed in both decentralized estimation and control algorithms.

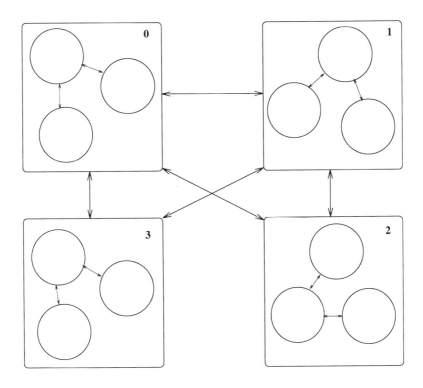

**FIGURE 8.1**
**Four WMRs Coupled by Information**

## 8.4.2 Applications

More applications to modular robotics can be accomplished. For WMRs, bigger vehicles with over ten active units could be designed, constructed and tested. This will clearly demonstrate the scalability and generality of the techniques developed. In order to provide information for navigation and WMR position estimation, more sonar sensors and perhaps an inertial navigation system (INS) could be mounted on the vehicle. The performance of these sensors and the associated sensor fusion could be improved to ensure maximum use of captured information. For high speed navigation, a dynamic WMR model could be derived.

A further application would be the control of a fleet of uncoupled, communicating WMRs. These WMRs would be independent of each and only linked by communication signals. Their internal structure would be modular and non-fully connected and the fleet network would be non-fully connected. Figure 8.1 illustrates this application to a fleet of four WMRs.

The modular design philosophy and the decentralized estimation and control can also be applied to vehicles such as the Mars Sojourner Rover with dramatic improvement of the vehicle's performance, competence, reliability and survivability. Similarly, other multisensor robotic systems such as the MIT Humanoid Robot (Cog) are potential applications.

The application to wheeled mobile robotics chosen in this book attests to what was then the main research interest at Oxford University. However, the variety of other possible applications is broad, including such fields as space structures, in particular the design and construction of shaped sensors and actuators, flexible manufacturing structures and financial forecasting. The general class of large scale systems provides the bulk of the potential applications which include air traffic control, process control of large plants, the Mir Space Station and space shuttles such as Columbia.

# Bibliography

[1] J. Abidi and S. Gonzalez. *Data Fusion In Robotics and Machine Intelligence*. Academic Press, 1992.

[2] M.D. Adams, P.J. Probert, and H. Hu. Toward a real-time architecture for obstacle avoidance in mobile robotics. In *Proc. IEEE Int. Conf. Robotics and Automation*, 1990.

[3] J.K. Aggarwal. *Multisensor Fusion for Computer Vision. NATO ASI Series*. Springer-Verlag, 1993.

[4] J.C. Alexander and J.H. Maddocks. *On the Kinematics of Wheeled Mobile Robots*. Springer-Verlag, 1988.

[5] J.C. Alexander and J.H. Maddocks. On the maneuvering of vehicles. In *SIAM Journal*, pages 48(1):38–51, 1988.

[6] A. T. Alouani. Nonlinear data fusion. In *Proc. 29th CDC*, pages 569–572, Tampa, December 1989.

[7] B.D.O. Anderson and J.B. Moore. *Optimal Filtering*. Prentice Hall, 1979.

[8] M. Athans. Command and control theory: A challenge to control science. *IEEE Trans. Automatic Control*, 32(4):286–293, 1987.

[9] M. Bacharach. Normal bayesian dialogues. *J. American Statistical Soc.*, 74:837–846, 1979.

[10] Y. Bar-Shalom. Tracking methods in a multi-target environment. *IEEE Trans. Automatic Control*, 23(4):618–626, 1978.

[11] Y. Bar-Shalom. On the track to track correlation problem. *IEEE Trans. Automatic Control*, 25(8):802–807, 1981.

[12] Y. Bar-Shalom and T.E. Fortmann. *Tracking and Data Association*. Academic Press, 1988.

[13]  Y. Bar-Shalom and X. Li. *Estimation and Tracking*. Artech House, 1993.

[14]  Y. Bar-Shalom and X. Li. *Multitarget-Multisensor Tracking*. YBS, 1995.

[15]  B.J. Barnett and C. D. Wickens. Display Proximity in Multicue Information Integration. *Human Factors*, 30(1):15–24, 1988.

[16]  S. Barnett and R.G. Cameron. *Introduction to Mathematical Control Theory*. Oxford Applied Mathematics and Computing Science, 1985.

[17]  R.E. Belman. *Dynamic Programming*. Princeton University Press, 1957.

[18]  R.E. Belman and S.E. Dreyfus. *Applied Dynamic Programming*. Princeton University Press, 1962.

[19]  R.E. Belman and R.E. Kalaba. *Dynamic Programming and Modern Control Theory*. Academic Press, 1965.

[20]  T. Berg. *Model Distribution in Decentralized Multisensor Data Fusion*. PhD thesis, Oxford University U.K., 1993.

[21]  T. Berg and H.F. Durrant-Whyte. Model distribution in decentralized sensing. Technical Report 1868/90, Oxford University Robotics Research Group, 1990.

[22]  T. Berg and H.F. Durrant-Whyte. Model distribution in decentralized multi-sensor fusion. In *Proc. American Control Conference*, pages 2292–2294, 1991.

[23]  J.O. Berger. A robust generalized bayes estimator and confidence region for a multivariate normal mean. *The Annals of Statistics*, 8:716, 1980.

[24]  J.O. Berger. *Statistical Decision Theory (second edition)*. Springer Verlag, Berlin, GDR, 1985.

[25]  S.S. Blackman. *Multiple Target Tracking with Applications to Radar*. Artech House, 1986.

[26]  D.A. Bradley and D. Dawson. *Mechatronics: Electronics in Products and Processes*. Chapman and Hall, 1996.

[27]  J.M. Brady, H.F. Durrant-Whyte, H. Hu, J. Leonard, P. Probert, and B.S. Rao. Sensor based control of AGVs. *IEE Computing and Control Journal*, 1(1):64–71, 1990.

[28]  R. Bronson. *Theory and Problems of Matrix Operations*. McGraw-Hill, 1989.

[29] R.A. Brooks. A layered intelligent control system for a mobile robot. In *Third Int. Symp. Robotics Research*, Gouvieux, France, 1986. MIT Press.

[30] R.A. Brooks. A robust layered control system for a mobile robot. *IEEE J. Robotics and Automation*, 2(1):14, 1986.

[31] R.A. Brooks. From earwigs to humans. *Robotics and Autonomous Systems (to appear)*, 1997.

[32] R.A. Brooks and L.A. Stein. Building brains for bodies. *MIT AI Lab Memo. number 1439*, 1993.

[33] T.P. Burke. *Design of a Wheeled Modular Robot*. PhD thesis, Oxford University, U.K., 1994.

[34] D.E. Catlin. *Estimation, control and the discrete Kalman filter*. Springer Verlag, 1989.

[35] K.C. Chang and Y. Bar-Shalom. Distributed adaptive estimation with probabilistic data association. *Automatica*, 25(3):359–369, 1989.

[36] K.C. Chang and Y. Barshalom. Distributed multiple model estimation. In *Proc. American Control Conference*, pages 866–869, 1987.

[37] K.C. Chang, C.Y. Chong, and Y. Bar-Shalom. Joint probabilistic data association in distributed sensor networks. *IEEE Trans. Automatic Control*, 31(10):889–897, 1986.

[38] C. Chong, S. Mori, and K. Chan. Distributed multitarget multisensor tracking. In Y. Bar-Shalom, editor, *Multitarget Multisensor Tracking*. Artech House, 1990.

[39] I.J. Cox and T. Blanche. Position estimation for an autonomous robot vehicle. In *IEEE/RSJ Int. Conf. on Intelligent Robot and Systems (IROS)*, pages 432–439, 1989.

[40] J. Crowley. World modeling and position estimation for a mobile robot using ultra-sonic ranging. In *Proc. IEEE Int. Conf. Robotics and Automation*, pages 674–681, 1989.

[41] H.F. Durrant-Whyte. Sensor models and multi-sensor integration. *Int. J. Robotics Research*, 7(6):97–113, 1988.

[42] H.F. Durrant-Whyte and B.S. Rao. A transputer-based architecture for multi-sensor data-fusion. In *Transputer/Occam Japan 3*, 1990.

[43] Editorial. The mars sojourner rover. In *IEEE Robotics and Automation Magazine*, pages 1–10, 1997.

[44] J. Fraden. *AIP Handbook of Modern Sensors*. AIP Press, 1995.

[45] B. Friedman. *Control System Design: An Introduction to State Space Methods.* McGraw-Hill, 1987.

[46] S. Grime. *Communication in Decentralized Sensing Architectures.* PhD thesis, Oxford University, U.K., 1992.

[47] S. Grime, H.F. Durrant-Whyte, and P. Ho. Communication in decentralized sensing. Technical Report 1900/91, Oxford University Robotics Research Group, 1991.

[48] S. Grime, H.F. Durrant-Whyte, and P. Ho. Communication in decentralized sensing. *Proc. American Control Conference, (ACC)*, 1992.

[49] G. Harp. *Transputer Applications.* Computer Systems Series, 1989.

[50] H.R. Hashemipour, S. Roy, and A.J. Laub. Decentralized structures for parallel kalman filtering. *IEEE Trans. Automatic Control,* 33(1):88–93, 1988.

[51] T. Henderson. Workshop on multi-sensor integration. Technical Report UUCS-87-006, Utah University, Computer Science, 1987.

[52] Hewlett-Packard. *The HCTL-1100 Controller.* 1991.

[53] Hewlett-Packard. *The HEDS-5500 Encoder.* 1991.

[54] P. Ho. *Organization in Distributed Systems.* PhD thesis, Oxford University, U.K., 1996.

[55] Y.C. Ho. Team decision theory and information structures. *Proceedings of the IEEE,* 68:644, 1980.

[56] Y.C. Ho and K.C. Chu. Team decision theory and information structures in optimal control. *IEEE Trans. Automatic Control,* 17:15, 1972.

[57] Y.C. Ho and S.K. Mitter. *Directions in Large Scale Systems.* Plenum Press, 1975.

[58] C.A.R. Hoare. *Communicating Sequential Processes.* Prentice Hall, 1985.

[59] H. Hu and P.J. Probert. Distributed architectures for sensing and control in obstacle avoidance for autonomous vehicles. In *IARP Int. Conf. Multi-Sensor Data Fusion,* 1989.

[60] S. Julier. *Process Models for the Navigation of High-Speed Land Vehicles.* PhD thesis, Oxford University, U.K., 1997.

[61] S. Julier and J.K. Uhlmann. A nondivergent estimation algorithm in the presence of unknown correlations. In *Proc. American Control Conference,* pages 656–660, 1997.

[62] S. Julier, J.K. Uhlmann, and H.F. Durrant-Whyte. A new approach for filtering nonlinear systems. In *Proc. American Control Conference*, pages 229–678, 1995.

[63] R.E. Kalman. A new approach to linear filtering and prediction problem. *ASME J. Basic Engineering*, 82:35–45, 1960.

[64] R.E. Kalman and R.S. Bucy. New results in linear filtering and prediction theory. *ASME J. Basic Engineering*, 83:95–108, 1961.

[65] R.D. Klafter and T.A. Chmielewski. *Robotic Engineering: An Integrated Approach.* Prentice Hall 1989, 1988.

[66] P. Lancaster and M. Tismenetsky. *The Theory of Matrices, Second Edition with Applications.* Academic Press, 1985.

[67] J.J. Leonard and H.F. Durrant-Whyte. Simultaneous map building and localization for an autonomous mobile robot. In *IEEE Int. Conf. on Intelligent Robot Systems (IROS)*, pages 1442–1447, 1991.

[68] R.C. Luo. Data fusion and sensor integration: State of the art in the 1990s. In *Data Fusion In Robotics And Machine Intelligence*, pages 7–136. Academic Press, 1992.

[69] R.C. Luo and M.G. Kay. Multisensor integration and fusion in intelligent systems. *IEEE Trans. Systems Man and Cybernetics*, 19(5):901–931, 1989.

[70] J. Manyika and H.F. Durrant-Whyte. *Data Fusion and Sensor Management: A Decentralized Information-Theoretic Approach.* Ellis Horwood Series, 1993.

[71] J. Manyika, S. Grime, and H.F. Durrant-Whyte. A formally specified decentralized architecture for multi-sensor data fusion. In *Transputing '91*, pages 609–628. IOS Press, 1991.

[72] Y. Matsuoka. *Embodiment and Manipulation Learning Process for a Humanoid Robot.* MS thesis, Massachusetts Institute of Technology, EECS, 1995.

[73] L. Matthies, E. Gat, R. Harrison, B. Wilcox, R. Volpe, and T. Litwin. Mars microrover navigation: performance evaluation and enhancement. *Autonomous Robots*, 2(4):291–311, 1995.

[74] P.S. Maybeck. *Stochastic Models, Estimation and Control, Vol. I.* Academic Press, 1979.

[75] P.S. Maybeck. *Stochastic Models, Estimation and Control, Vol. 2.* Academic Press, 1982.

[76] P.S. Maybeck. *Stochastic Models, Estimation and Control, Vol. 3.* Academic Press, 1982.

[77] R. McKendall and M. Mintz. Data fusion techniques using robust statistics. In *Data Fusion in Robotics and Machine Intelligence*, pages 211–244. Academic Press, 1992.

[78] P.F. Muir. *Modeling and Control of Wheeled Mobile Robots*. PhD thesis, Carnegie Mellon University, U.S.A., 1988.

[79] P.F. Muir and C.P. Neuman. Kinematic modeling of wheeled mobile robots. In *Journal of Robotic Systems*, pages 4(2):281–33, 1987.

[80] A.G.O. Mutambara. *A Formally Verified Modular Decentralized Control System*. MSc thesis, Oxford University, U.K., 1992.

[81] A.G.O. Mutambara. *Decentralized Estimation and Control with Applications to a Modular Robot*. PhD thesis, Oxford University, U.K., 1995.

[82] A.G.O. Mutambara. Decentralized data fusion and control for a mobile robot. In *Florida Conference on Recent Advances in Robotics*, pages 163–181, 1996.

[83] A.G.O. Mutambara. Nonlinear sensor fusion for a mobile robot. In *SPIE's International Symposium on Intelligent Systems and Advanced Manufacturing; Sensor Fusion and Distributed Robotic Agents*, pages 2905–11:102–113, 1996.

[84] A.G.O. Mutambara. Fully connected decentralized estimation. In *SPIE's International Symposium on Photonics for Industrial Applications, Sensor Fusion VII*, pages 3209–03:123–134, 1997.

[85] A.G.O. Mutambara. Fully connected decentralized estimation: A robotics application. In *Florida Conference on Recent Advances in Robotics*, pages 151–159, 1997.

[86] A.G.O. Mutambara. A modular wheeled mobile robot. *Journal of Microcomputer Applications (to appear)*, 1998.

[87] A.G.O. Mutambara and H.F. Durrant-Whyte. A formally verified modular decentralized robot control system. In *IEEE/RSJ Int. Conf. on Intelligent Robot and Systems (IROS)*, pages 2023–2030, 1993.

[88] A.G.O. Mutambara and H.F. Durrant-Whyte. Modular decentralized robot control. In *Intelligent Vehicles Symposium*, pages 512–518, 1993.

[89] A.G.O. Mutambara and H.F. Durrant-Whyte. Distributed decentralized robot control. In *American Control Conference*, pages 2266–2267, 1994.

[90] A.G.O. Mutambara and H.F. Durrant-Whyte. Modular scalable robot control. In *IEEE Conf. on Multisensor Fusion Integration*, pages 512–518, 1994.

[91] A.G.O. Mutambara and H.F. Durrant-Whyte. Nonlinear information space: A practical basis for decentralization. In *SPIE's International Symposium on Photonics for Industrial Applications, Sensor Fusion VII*, pages 2355, 10:84–95, 1994.

[92] A.G.O. Mutambara and H.F. Durrant-Whyte. The decentralized extended information filter. *Automatica, A Journal of IFAC (to appear)*, 1998.

[93] A.G.O. Mutambara and H.F. Durrant-Whyte. Estimation and control for a modular wheeled mobile robot. *The International Journal of Robotics (to appear)*, 1998.

[94] A.G.O. Mutambara and M.Y Haik. State and information estimation: A comparison. In *American Control Conference*, pages 2266–2267, 1997.

[95] A.G.O. Mutambara and M.Y. Haik. State and information estimation for linear and nonlinear systems. *ASME Journal of Dynamic Systems, Measurement and Control*, 1997.

[96] A.G.O. Mutambara and M.Y. Haik. EKF based parameter estimation for a heat exchanger. *ASME Journal of Dynamic Systems, Measurement and Control (to appear)*, 1998.

[97] N. Nandhakumar and J. Aggarwal. Integrating information from thermal and visual images for scene analysis. In *Proc. SPIE Conf. Applications of Artificial Intelligence*, pages 132–142, 1986.

[98] NASA. *Space Shuttle Mission STS-87 Press Kit*. NASA Space Transportation System, U.S.A., 1997.

[99] NASA. *Space Shuttle Reference Manual*. NASA Space Transportation System, U.S.A., 1997.

[100] NASA. *STS-86 Press Information and Mission Time Line*. Boeing Reusable Launch Systems, PUB 3546-V Rev 9-97, MTD 970918-6447, 1997.

[101] W.L. Nelson and I.J. Cox. Local path control for an autonomous vehicle. In *Proc. IEEE Int. Conf. Robotics and Automation*, pages 1504–1510, 1988.

[102] R. Penrose. A generalized inverse for matrices. In *Proc. Cambridge Phil. Soc.*, pages 51:406–413, 1955.

[103] R.M. Pringle and A.A. Rayner. *Generalized Inverse Matrices With Applications to Statistics.* Charles Griffen and Company Limited, 1971.

[104] T. Queeney. Generic architecture for real-time multisensor fusion tracking algorithm development and evaluation. In *The International Society for Optical Engineering (SPIE)*, pages 345–350, 1994.

[105] B. Quine, J.K. Uhlmann, and H.F. Durrant-Whyte. Implicit Jacobians for linearized state estimation in nonlinear system. In *Proc. American Control Conference*, pages 559–564, 1995.

[106] B.S. Rao and H.F. Durrant-Whyte. A fully decentralized algorithm for multi-sensor kalman filtering. Technical Report 1787/89, Oxford University Robotics Research Group, 1989.

[107] B.S. Rao and H.F. Durrant-Whyte. A fully decentralized algorithm for multi-sensor kalman filtering. *IEE Transactions Schedule D*, 138(5):413–420, 1991.

[108] B.S. Rao, H.F. Durrant-Whyte, and A. Sheen. A fully decentralized multi-sensor system for tracking and surveillance. *Int. J. Robotics Research*, 1991.

[109] C.R. Rao and S.K. Mitra. *Generalized Inverse of Matrices and Its Applications.* John Wiley, 1971.

[110] D.B. Reid. An algorithm for tracking multiple targets. *IEEE Trans. Automatic Control*, 24(6), 1979.

[111] J.M. Richardson and K.A. Marsh. Fusion of multisensor data. *Int. J. Robotics Research*, 7(6):78–96, 1988.

[112] S. Roy and P. Mookerjee. Hierarchical estimation with reduced order local observers. In *In Proc. of the 28th Conf. on Decision and Control*, pages 420–428, 1989.

[113] N.R. Sandell, P. Varaiya, M. Athans, and M.G. Safonov. Survey of decentralized control methods for large scale systems. *IEEE Trans. Automatic Control*, 23(2):108–128, 1978.

[114] S. Shafer, A. Stenz, and C. Thorpe. An architecture for sensor fusion in a mobile robot. In *Proc. DARPA Workshop on BlackBoard Architectures for Robot Control*, 1986.

[115] D.D. Siljak. *Decentralized Control of Complex Systems.* Academic Press, 1991.

[116] J. F. Silverman and D. B. Cooper. Bayesian clustering for unsupervised estimation of surface and texture models. *IEEE Trans. Pattern Analysis and Machine Intelligence*, 10(4):482–495, 1988.

[117] J.L. Speyer. Communication and transmission requirements for a decentralized linear-quadratic-gaussian control problem. *IEEE Trans. Automatic Control*, 24(2):266–269, 1979.

[118] A. Stevens, M. Stevens, and H.F. Durrant-Whyte. Oxnav: Reliable autonomous navigation. In *IEEE International Conference on Robotics and Automation (ICRA)*, pages 2607–2612, 1995.

[119] G. Tadmor. Control of large discrete event systems, constructive algorithms. *IEEE Trans. Automatic Control*, 34(11):1164–1168, 1989.

[120] J.N. Tsitsiklis and M. Athans. On the complexity of decentralized decision-making and detection problems. *IEEE Trans. Automatic Control*, 30(5):440–446, 1985.

[121] T. Tsumura. Survey of automated guided vehicles in Japanese factories. In *Proc. IEEE Int. Conf. Robotics and Automation*, page 1329, 1986.

[122] H.S. Tzou, G.G. Wen, and C.I. Tseng. Dynamics and distributed vibration controls of flexible manipulators. In *Proc. IEEE Int. Conf. Robotics and Automation*, pages 1716–1725, 1988.

[123] J.K. Uhlmann. *Dynamic Map Building and Localization: New Theoretical Foundations*. PhD thesis, Oxford University, U.K., 1996.

[124] J.K. Uhlmann. General data fusion for estimates with unknown cross covariances. In *Proc. of the SPIE Aerosense Conference, Vol. 2755*, pages 536–547, 1996.

[125] B.W. Wah and G.J. Li. A survey on the design of multiprocessing architectures for artificial intelligence applications. *IEEE Trans. Systems Man and Cybernetics*, 19(4):667–693, 1989.

[126] E.L. Waltz and J. Llinas. *Sensor Fusion*. Artech House, 1991.

[127] M. Williamson. Postural primitives: Interactive behavior for a humanoid robot arm. *Presented at SAB, Cape Cod, MA*, 1996.

# Index